Northern Ireland and Beyond

The GeoJournal Library

Volume 33

The titles published in this series are listed at the end of this volume.

Northern Ireland and Beyond

Social and Geographical Issues

by

EMILIO BIAGINI

Department of Economic and Social Research,
University of Cagliari, Italy

KLUWER ACADEMIC PUBLISHERS
DORDRECHT / BOSTON / LONDON

Library of Congress Cataloging-in-Publication Data

Biagini, Emilio.
    Northern Ireland and beyond / Emilio Biagini.
        p.   cm. -- (GeoJournal library ; v. 33)
    Includes bibliographical references.
    ISBN 0-7923-4046-9 (alk. paper)
    1. Northern Ireland--Historical geography. 2. Northern Ireland-
    -History.   I. Title.   II. Series.
    DA990.U46B39   1996
    941.6--dc20   96-11231

ISBN 0-7923-4046-9

Published by Kluwer Academic Publishers,
P.O. Box 17, 3300 AA Dordrecht, The Netherlands.

Kluwer Academic Publishers incorporates
the publishing programmes of
D. Reidel, Martinus Nijhoff, Dr W. Junk and MTP Press.

Sold and distributed in the U.S.A. and Canada
by Kluwer Academic Publishers,
101 Philip Drive, Norwell, MA 02061, U.S.A.

In all other countries, sold and distributed
by Kluwer Academic Publishers Group,
P.O. Box 322, 3300 AH Dordrecht, The Netherlands.

Figures and photos not otherwise credited, are copyright of the author.

Printed on acid-free paper

Printed in the Netherlands

CONTENTS

FOREWORD

The Celts were the first peoples to experience as Anglo-Saxon invasion, beginning with the Britons of the Province of Brittania after the withdrawal of the Romans; then what was left of them was also subject to attack in the so-called "Celtic periphery", including Ireland, after the conquest of what is now England was consolidated.

Having built their empire in the British Isles, the English moved overseas and made their power felt in all continents. Where they did not conquer, or where their conquest did not last, they managed to exercise a deep cultural influence.

In this way English has become *the* language of the modern world, as Latin and Greek had become the languages of the ancient Euro-Mediterranean world.

There is something grand and admirable in all that, in the power and civilization of empires like the Roman and the British. But there is also something inherently vain in every human endeavour: empire building is by no means the last vain.

After the First World War, the British Empire was seemingly as strong as ever and, but for the loss, a long time before, of a part of North America, still largely intact.

Yet the colonial peoples did not longer accept to be ruled indefinitely.

The first blow, however, came from very near home: the break-up of the political unity of the British Isles by the unruly Irish: and the last blow is likely to come from Ireland too, after all other colonies have gone.

Northern Ireland is a fragment of a lost empire. The natives there were subdued but never assimilated, whilst the descendants of the settlers have been strong enough to oppose reunification so far, but are unable either to turn the tide or to get adjusted to a united Ireland.

So, in a way, this is a book on vanity, the vanity of the human condition. The vanity of power and conquest; the vanity of resisting inevitable change; the vanity of hatred and violence from any quarter whatsoever; the vanity of victory, as the victors too are destined to be ground to dust.

LIST OF FIGURES

Scotland in 1707. By "British", in this case, is meant either an English or a Scottish settler.

3.1 — Population growth and percentages of Catholics by Province. (Sources: for 1732 *An Abstract of the Numbers of Protestant and Popish Families in the several Counties and Provinces of Ireland taken from the returns made by the Hearthmoney Office in Dublin, 1723-33*, Dublin, 1734; for 1834 *First Report of the Commission on Public Instruction, Ireland*, London, 1835, completed by data of Beaumont, 1839 and McCulloch, 1839).

3.2 - Social conditions in the Irish countryside from the *Third Report of the Irish Poor Inquiry Commissioners*, London, H.M. Stationery Office, 1837, and the *Report from Her Majesty's Commissioners of Inquiry into the state of the law and practice in respect to the occupation of land in Ireland*, 3 vols., Dublin, H.M. Stationery Office, 1845.

3.3 — Two Irish villages, in the second half of the nineteenth century, looked totally different: (i) Annvale, Co. Armagh, in Ulster, and (ii) Tullig, Co. Kerry, in Munster. They are good examples of the contrast between the comparative prosperity of better accessible and more industrialized areas and the depopulation of peripheral areas in the aftermath of the Great Famine (by kind permission of Oxford University Press).

4.1 — Social pyramids in Ulster and in the other three provinces at the time of partition (1921). In the event, partition did not occur on provincial lines, but Ulster was also partitioned, so that the common usage of the name "Ulster" to indicate Northern Ireland, which has only 6 counties, is utterly wrong.

5.1 — Relationships between central and peripheral areas compared with local majorities. Although based upon old data, the pattern shown here still holds. If anything, segregation has tended to increase. The maps are based upon a subdivision of present-day Northern Ireland into 67 local administrative units (Source: Hoare).

5.2 — First component (accounting for 39.9% of total variance) of principal components analysis of economic and social imbalances in Northern Ireland. Explanation in text.

6.1 — Regional planning framework.

7.1 — First component (accounting for 54.5% of total variance) of principal components analysis of social pathologies in the urban area of Belfast. Explanation in text.

8.1 — A model of the conflict process in a traditional society.

8.2 — Conditions paving the way to conflict after the onset of the industrial revolution.

8.3 — A model of controlled conflict under an external superordinate authority.

LIST OF PHOTOS

1 — The Giant's Causeway, on the northern coast of Co. Antrim.

2 — The columnar basalts of the Giant's Causeway exploited as a tourist attraction.

3 — Carrickfergus Castle (late 12th - 13th century). A control point at a gateway of colonial penetration. In the early 17th century it became one of the main power bases of the Chichesters and a grim jail for Sir Con O'Neil of Clandeboy, whose lands were conveted by the invaders (Department of the Environment for Northern Ireland. Crown Copyright).

4 — Stormont Building: formerly a sign of Unionist power.

5 — An Orange Order march. The still picture does not unfortunatemly convey the loud noise which is an essential part of the show.

6 — Loyalist simbols in South Belfast: a mark of territorial control.

7 — Lower Falls, West Belfast.

8 — A gutted building in South Belfast: the effect of an incendiary device planted in the shop at the ground floor by the IRA.

9 — A civil war landscape in Central Belfast: the effect of an IRA car bomb.

10 — One more building in Central Belfast razed by an IRA car bomb.

11 — Security barrier erected as a protection of the central business district of Belfast.

12 — The wall parting two ghettoes in Central Belfast: Catholic Falls and Protestant Shankill (From *De Telegraaf*, Amsterdam. Copyright Reuter).

INTRODUCTION

Northern Ireland came into existence, as a region of the United Kingdom, in 1921. When the greater part of Ireland attained independence, the Protestant-Unionist population of six out of the thirty-two Irish counties managed to retain the old British connection in spite of the presence of a very substantial Catholic minority who might have preferred to live under an Irish government.

This partition brought forth a true problem region, with an area of 14,000 square kilometres only — one sixth of the whole of Ireland, and considerably smaller than Wales. It stands to reason that Northern Ireland cannot be studied in isolation. Reference will have to be made to Ireland as a whole, to Eire, to Great Britain and to interactions of Ireland with the rest of the world, especially North America.

The consequences, economic and political, of the partition are still with us as Europeans, and are worth looking at. To understand the consequences, however, it is necessary to know the causes. Any attempt to grasp a current situation calls for a scrutiny of the beginnings, the historical roots of present problems, as authoritatively suggested by Tocqueville. We cannot trust "instant books" trying to take snapshots of the latest facts and sometimes seeking to interpret them without considering how they evolved. Another snag to be studiously avoided, in my opinion, is following the fashions of the "politically correct" from any quarter soever; for we must not seek the favour of the world, but the truth.

Books on Northern Ireland already fill a great many bookshelves. Its problems have been approached by many authors from all angles. But they have not yet been solved, and until a solution is forthcoming, there will be something to say; perhaps, from time to time, even something new. When a solution is found, someone will come up with something to say on, or against, the solution itself, and so the publishers (and, hopefully, the readers) will still be kept busy.

I have been working on Northern Ireland since 1984, and being neither Irish nor British, I may perhaps have been able to see things under a somewhat impartial angle. And yet, as a foreigner? Is not living in a common European homeland a sufficient reason for a concern about what happens in any part of our own home, Europe?

The approach of this book will not please some. If "to please" is meant "to attempt to curry favour," I have sought to "please" no one. I have trespassed on entrenched prejudices from different quarters. I consistently call Derry by its original name, rather than "Londonderry," thus angrying the Unionists. I call the British Isles with their proper name because the refusal of this name by the extremist Irish Republicans is ludicrous: were not the Britons Celt, and not Anglo-Saxon, after all? By choosing the names "Great Britain" and "British Isles," the English have paid homage, whether wittingly or not, to the Celtic past. So I will probably make the Republicans angry too. No apology is offered for these and similar cases.

Understanding a complex situation may assist a little in changing it. And if Western Europeans, and Westerners in general, cannot solve their own ethnic conflicts, how can they help others in the savage revival of ethnic and tribal struggles now spreading all over the world?

A short explanation of the title of this book is in order. The gist of the argument is that history is moving *beyond* Northern Ireland: a politically artificial construction which the British government of the day did not even particularly want. Only the Protestant-Unionists, with the support of British Conservatives, wanted it, and forced the rest of the world to accept it. Eventually, their regime collapsed, compelling Westminster to establish direct rule over the region. In time — we do not yet know when and at what cost — a united Ireland seems inevitable, and it is in the best interest of Britain to help in achieving it.

Some scholars will find fault with this book as I have found fault with theirs. This is part of the unavoidable clash of views and interests in a harsh and long-drawn ethnic struggle. A study on conflict cannot help becoming itself a part of the conflict and attract criticism from those who perceive their interests to be under attack: it is a perfectly normal consequence of any honest quest for truth.

This my second book on Ireland. The first one, in Italian, published in 1992, deals with Ireland as a whole. The title is *Irlanda. Sviluppo e conflitto alla periferia d'Europa*, which translates in English to: *Ireland. Development and conflict at the periphery of Europe*. I must say with some satisfaction that I have been encouraged to translate the book by a review in *Irish Geography*.

Instead of a translation, I thought it best to produce a new book. This time I have been led to focus only upon Northern Ireland by the fact that the subject appears to be of more vibrant and controversial interest — controversies are the very heart of the quest for truth — due to the ongoing conflict in the region. This book is not a mere reshuffling of some parts of the earlier book, as it contains a lot of new material and has been considerably updated.

As I am an academic (University of Cagliari, Italy), I tend to inflict my books on my luckless students. So, among other things, they have to study Ireland. They — my students — seem rather interested, perhaps because the insular and rather peripheral situation of the "emerald isle" calls for intriguing comparisons with our own beautiful island of Sardinia. A more disenchanted view is that they must study it or fail the examination.

I have, unfortunately for myself and luckily for my English-speaking readers, no such leverage upon them. I will therefore, endeavour not to be too heavy, which is a rather hard task for a Fellow (with the capital "F," meaning a Fellow of the Royal Geographical Society) weighing over fourteen stones.

My most sincere thanks go to my wife Maria Antonietta for her patience and to Prof. Russell King, Prof. Frederick Boal and Dr. Philip Robinson for encouragement and constructive criticism. Discussion with Prof. Paul Compton has also been most enlightening, though he, and the colleagues of his (Unionist) persuasion, will hardly be prepared to agree with the stance taken here. Of course, this stance, as well as any remaining inaccuracies, are my sole responsibility.

CHAPTER 1

A PHYSICAL SPACE
POWERFULLY SHAPED BY MAN

Isolation and a peripheral situation are two typical features of the Irish environment, both from a physical and a social viewpoint. Great Britain itself, though far larger (229,897 square kilometres in comparison with 84,403 of Ireland) and better accessible from the continent, existed in a somewhat peripheral condition until the Discovery of America.

Ireland was unable to profit from the revolutionary changes in the world maritime routes wrought by Christopher Columbus. It was rife with internal division and already under English egemony, which was later to become absolute domination. Finally, ships usually sailed from British ports straight to the New World, without any need to call at an Irish port.

But Ireland is by no means utterly isolated: it is extremely easy to penetrate from all sides, and quite close to Britain. The two islands seem to be "looking at each other," and are in fact mutually visible. About twenty kilometres separate Fair Head in the north of Co. Antrim and the Scottish peninsula of Kintyre. Britain and Ireland stand on the same oceanic platform and share a great deal of their geological histories.

The relationship is particularly close between Ulster and Scotland. In the Ordovician period, about 500 million years ago, both areas were a part of palaeo-North America, while the rock formations which would later be a part of England, Wales and southern and eastern Ireland belonged to the Eurasian plate and were divided by an ocean which geologists have chosen to call Iapetus (or palaeo-Atlantic).

The Iapetus shrunk as its oceanic crust was consumed by subduction under the Eurasian plate. As a consequence of this process, about the end of the Silurian, roughly 400 milion years ago, Scotland joined England along the so-called Iapetus suture, while the area that is now Ulster (and Connacht) became welded likewise to the rest of Ireland. The attendant tectonic compression brought about the Caledonian orogenesis, tending north-

1

east/south-west, whose worn-out relief is recognizable throughout Scotland as well as in northern and western Ireland.

When the Atlantic opened up again, from the late Mesozoic and especially during the Tertiary age (the boundary between the two ages being about 65 million years ago), considerable tension within the crust unleashed a spate of volcanic activity which brought forth the extensive covers of early Tertiary (Eocene) basaltic lava so typical of south-west mainland Scotland, the Hebrides and north-east Ireland.

Yet the great faults which bound Lowland Scotland are blurred in Ulster, at least to some extent, by formations of various ages, from Carboniferous to Tertiary, which overlap the dominant trend of the Caledonian orogeny.

In general, the solid geology of Ireland is dominated by Palaeozoic rocks, with important Archaeozoic outcrops, especially in the north-west, and volcanic ones (intrusive and effusive) particularly in Ulster. This northern part of Ireland is noted for its geological complexity: all main periods being represented, with the exception of the Cambrian (Fig. 1.1).

Ulster's basal rocks are metamorphic, Moinian and Dalradian in age (from about 1000 to 570 million years ago): they can be found outcropping in north-east Antrim, the Sperrin Mountains and Donegal. South of the Lagan valley, Ordovician and Silurian rocks (510 to 400 milion years ago) strike south-west into Monaghan and Cavan. Old Red Sandstone and Carboniferous rocks (400 to 290 million years ago) survive, after eons of erosion, only in sparse outliers.

Soft sedimentary deposits covered most of the area during the Mesozoic age (230 to 65 million years ago). In the north-east, the Mesozoic system was overlain by early Tertiary basalt lavas whose cooling process gave rise to peculiar columnar patterns so prominent in the Scottish island of Staffa (Inner Hebrides) and in the Giant's Causeway in north Antrim (Photo 1).

The antiquity of many Irish geological formations might raise hopes for rich mineral deposits, destined, however, to end largely in disappointment — an outstanding instance being the Carboniferous outcrops. These are very widespread, but represented mainly by limestone devoid of coal seams, since, during that period, the sea was deeper in the area which was to become Ireland, whereas coal seams were formed in coastal lagoon environments. Thus, when the Industrial Revolution started in Britain, fuelled by the plentiful coal seams, Ireland was largely bypassed by the new developments. The Irish economic predicament is thus not entirely due to policies of past British governments: to ignore the constraints imposed by natural patterns and processes would be an undue forcing of the evidence.

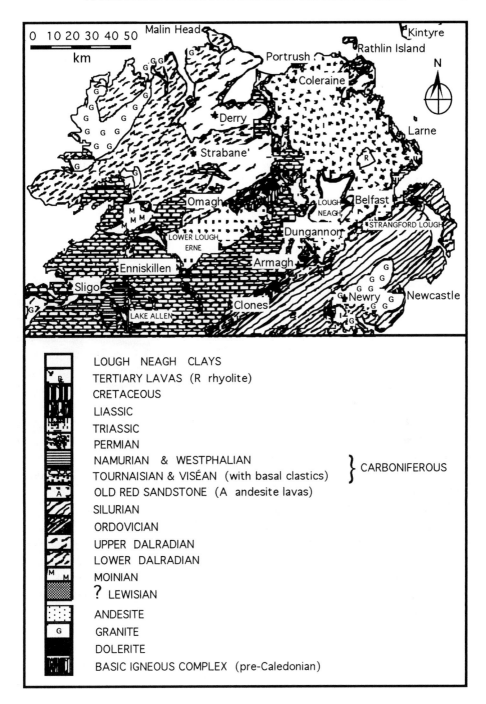

Fig. 1.1 — Sketch of the solid geology of Ulster (Source: Wilson).

Photo 1 — The Giant's Causeway on the northern coast of Co. Antrim.

A variety of mineral resources have been exploited, but have never become very important for a modern industrial development. Some iron furnaces were active in the seventeenth century, but most of them subsequently closed down due to exhaustion of timber and British competition. Copper, lead and some coal were mined here and there in the last century, but very little of this modest activity took place in Ulster. Since the Sixties of this century, a more extensive exploitation of lead, tin, silver, mercury, copper, barite, marble and peat has taken place in Eire.

Very little of that is to be found in Northern Ireland which is known to possesses almost exclusively building materials, peat and some lignite; in any case mineral exploitation is restricted there, mainly for environmental reasons. An exception to this is the gold bed in the Sperrin Mountains, close to the little town of Gortin. Considerable capital investment is needed to make it profitable; in 1987 it was being prospected with machines only (it being virtually impossible to obtain explosives in Northern Ireland due to the strict security measures) by Ennex International, whose headquarters are in Dublin. This a fine example of technical expertise from the Republic being employed in the development of the North.

The marine geology around Ireland is made up of tectonic basins filled with Mesozoic and Tertiary sediments where salt and coal can be found, besides the oil and natural gas nowadays widely exploited in Eire. Very few of these resources, however, have been found, so far, in Northern Irish waters.

Ulster was totally covered several times by Quaternary glaciations: the eastern half was usually invaded by ice from Scotland, while in the west, local centres of glaciations were dominant from Donegal and possibly from the Perrins Mountains. The traditional view (derived from early field studies in the Alps) of four glacial maxima is now superseded by a different model based upon the oxygen isotope record from undisturbed ocean sediments, thought to reflect major changes in global ice volume. The new picture entails about thirty alternate cold and temperate stages during the last 2.4 million years in the British Isles.

Pleistocene ice left strong marks in Ulster. Most of the traces date back, of course, to the last glaciation, each of the previous ones having being largely destroyed by atmospheric action and by the remodelling action of the following glacial event.

Among the erosional features left by the ice, the U-shaped morphology of vallies is very much in evidence. Other prominent features are glacial cirques or *corries* (e.g. on the north face of Slieve Corragh): hollows open downstream but bounded upstream by a rounded headwall. Extensive striations on hard rocks and the relics of glacial lakes can also be seen.

Glacial depositional features include extensive morainic drifts which blanketed the whole territory except the highest hills, and nowadays strongly affect settlement patterns and agriculture. Such drift geology also obscures the underlying solid rock geology.

Prominent depositional features left by the ice are *kames, eskers* and *drumlins.*

Kames are rounded heaps of gravel and sand, usually no more than 10-15 metres high, probably formed by the deposition of surface moraine when the ice cap melted. They are often accompanied by cavities left by thickenings in the ice sheet, presently occupied by small lakes or marshes.

Eskers are gravelly beds made up by bottom glacial drift compressed by the ice mass: perhaps beds of glacial streams flowing under the ice.

Drumlins are oval-shaped heaps of sand and gravel, usually up to 500 metres long and up to 30 metres high, with a rounded and a pointed end, the latter in the direction towards which the ice flowed; the faster the ice was moving, the more elongated they are. Drumlins are always in swarms of dozens or even hundreds. In the northern part of Ireland, a distinct drumlin belt has been identified (Fig. 1.2). Off the coasts, partially submerged drumlins give rise to typical islets.

Glaciations have thus contributed to smoothing the land, whose relief, dating back, as we have seen, to the Caledonian orogenic uplift, is now ground down to mere hills (Fig. 1.3). These are not part of one single system but scattered close to the coasts.

The coasts are often high, with narrow deep bays not unlike fjords (*loughs*). Numerous raised beaches witness changing sea levels and crustal displacements due to the rebounding of the crust after each glacial episode. Glaciomarine deposits indicating sea levels up to 80 to 90 metres higher than at present, dating between 16,000 to 15,000 BC, have been identified in Co. Antrim (near Ballycastle and at Portballintrae) and Co. Donegal (Malin Beg). Several fragmentary late glacial shorelins have been identified up to a height of 30 metres; they have mostly been dated to the late Midlandian deglaciation (about 15,000 to 13,000 BC). Around 10,000 to 8,000 BC, sea levels dropped by 100 or even 120 metres. The subsequent rise (marine transgression) was very swift, recrossing the present shoreline by 4500 BC, and rising above the present sea level by a few metres as shown by evidence from Carnlough (Co. Antrim) and the mouth of the Bann. It was probably during that last rise that Ireland finally became an island. The better preserved raised beaches reach heights up to 5 metres and are much

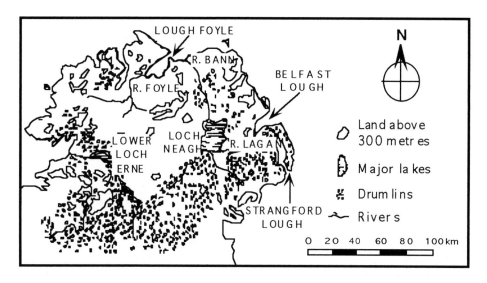

Fig. 1.2 — Geomorphology: drumlin belts and hydrology (Source: Orme).

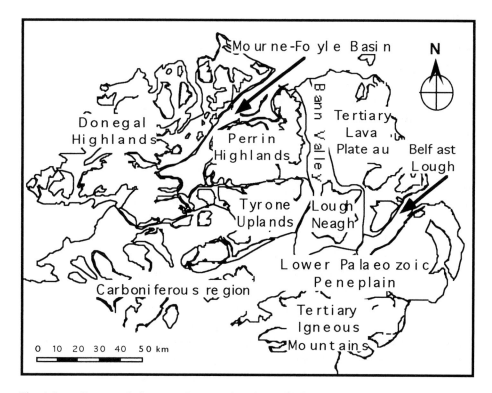

Fig. 1.3 — Geomorphology: main natural regions of Ulster (Source: Wilson).

younger, dating back to the mid-Holocene (about 4000 BC).

The climate history of Ireland in late-glacial and "post-glacial" times can be traced in the pollen record preserved in peat layers, and also in the fossil record of mammals and molluscs whose range of climatic toleration is known, at least to some extent.

The melting of the last major ice sheet began about 16,000 BC. The Holocene (i.e. the last 10,000 years) is the present geological period and is regarded by some as "postglacial." There is no reason why it should not, instead be merely an interglacial, swiftly followed (in a mere few thousand years) by a new glaciation.

In the wake of the disappearing ice-sheets, tundra began at first to recolonize the land: there was an abundance of the Dwarf Willow (*Salix herbacea*). Solifluxion deposits formed on hill slopes, while in fluvio-glacial sand and gravel deposits ground ice wedge pseudomorph structures developed, notably in the Sperrin and Co. Derry, with cryoturbation structures commonly associated. A milder interval, known as the Allerød period, is characterized by the widespread occurrence of the now extinct Giant Irish Deer (*Megaceros giganteus*).

With rising temperatures, hazel, alder, pine, elm and oak arrived. The Boreal climate (7000 to 5500 BC), was probably 2.5 °C warmer than today. When the Flandrian transgression (around 5500 BC) inundated the coast, peat were deposited in estuaries, along with grey sand, mud and silt. Estuarine clays occur in the Lagan estuary, Larne Lough, Strangford Lough and Lough Foyle, and contain a plentiful fauna of molluscs and foraminifera. Beneath Belfast, these unconsolidated clays are up to 17 metres thick. Their very low bearing strength causes serious problems in the central city area where all the substantial buildings must rest on piles.

In latter "post-glacial" times cooler and wetter climate encouraged the growth of blanket bog on the hills. Extensive areas of peat are exploited elsewhere in Ireland, whereas in Ulster, few areas have a sufficient volume of basin peat to justify large scale exploitation.

Blown sand formed considerable dunes, most of which are stable, in the Magilligan, Bannfoot, and Dundrum Bay areas, while landslips occured in chalk and basalt around the Antrim plateau and moved on the Lias shales and the block screes around the foot of Fair Head.

The coastal zone has been extensively altered by man. Only inaccessible cliffs or protected stretches such as the magnificent Eocenic columnar basalts of the Giant's Causeway have remained comparatively untouched, in spite of, sometimes, considerable pressure from tourism (Photo 2).

Photo 2 — The columnar basalts of the Giant's Causeway exploited as a tourist attraction.

Elsewhere, reclamation, draining, grazing, and recreational activities have greatly affected the original habitats. Coastal lagoons, very extensive in the past, have been drained, whilst the vegetation of sand dunes has been altered by agriculture and, more recently, by tourism.

A broad tectonic basin, perhaps subsiding since the Jurassic age, covered by Oligocenic clays, occupies the very centre of Ulster where the largest lake in the British Isles (396 square kilometres) is to be found: the Lough Neagh. It is not very deep, reaching only about 30 metres, and has rich fishing grounds with eels, salmon, trout and pollan (a fresh water fish akin to the herring). The Lough's shores used to be frequently flooded, but intensive reclamation has taken place from early in this century onwards, and such occurences are no longer so frequent, while the marshy areas around the lake have also become substantially narrower.

In their "natural" state, rivers were slow, uninterrupted by falls or rapids, due to a relief powerfully modelled by glacial activity underlying the drainage network. The abundance of navigable rivers and lakes eased the deep penetration of ships of small draught like those of the Vikings. If, together with the traditional political fragmentation, these drainage patterns contributed to make Ireland's territory open to invasions and hard to defend, it also made consolidation by a conquering army very difficult. Hidden in moors, marshes and woods, the defeated Irishmen were often able to prepare and unleash unexpected insurrections and reconquests.

Extensive man-made changes have been taking place lately in the river geomorphology, especially since EEC subsidies became available, thus making it possible to undertake drastic drainage and channelization works. This has reduced the likelihood of floods which used to be quite frequent, but has also entailed severe drawbacks. The natural drainage, with its sluggish flow, used to provide a great variety of habitats as every river behaved rather like a sequence of small lakes. Nowadays waters move more swiftly and are considerably warmer due to the cutting of the riverine vegetation. As a consequence, overall biomass and biodiversity have dropped dramatically. Not unexpectedly, this has severely affected the yield of the local salmon and trout fisheries, sometimes even cutting it by ninety per cent.

The comparatively small size of Ulster and its fragmentation into many river basins do not allow the formation of extensive rivers. The largest mean river discharge at the mouth is 113 cumecs (cubic metres per second), reached by the Bann which drains the Lough Neagh basin. Tributaries to the same lake have far lower mean discharge rates at their mouths into the lake: the Bann itself 17.9, the Blackwater 19.7, the Main 15.4. The Foyle discharges 58.4 cumecs at the mouth into the ocean, the Lagan 8.9. By way of comparison, the Shannon, the largest Irish river, which lies entirely outside Ulster, discharges at the mouth 178.5 cumecs.

The climate is typically oceanic, and particularly mild thanks to the influence of the Gulf Stream that touches the western and northern shores. Daily average temperatures almost never drop below 0 °C in the winter, while in the summer they seldom reach above 16°C. Differences between the coast and the interior are rather limited in this respect. Daily sunshine is severely restricted by frequent thick clouds borne by the westerlies (Fig. 1.4).

The weather is extremely variable being dominated in all seasons by frontal depressions from the North Atlantic. Of paramount importance is the maritime polar air, which originates above the Arctic and becomes milder and moisture-laden by passing over the ocean. In this way it becomes unstable, develops cumulus and cumulonimbus, and brings showers and thundershowers. Maritime tropical air, instead, originates from the Azores anticyclone and, as it moves north, is cooled by the ocean, and becomes saturated with humidity, forming stratus clouds with rain or drizzle. Airmasses from other directions of the compass are less frequent, though continental polar air from Siberia may reach Ulster in winter, bringing severe weather. Also in winter, an inflow of arctic air can sometimes occur, bringing unsettled weather and snow.

Precipitations are plentiful, there being no dry season: the months from August to January have averages between 150 and 140 mm, while the minimum is in April — slightly under 100 mm — thereafter they begin to rise again towards the autumn and winter peaks (Fig. 1.5). Geographically, precipitations are far from homogenous, being considerably higher in the west, due to the prevailing westerlies (Fig. 1.6). Potential evapotranspiration is higher on the coasts. Notwithstanding such differences, the climate lacks a marked regional diversification.

Although the lowlands, and even parts of the hills are covered by thick glacial drift, the soils usually reflect the character of the underlying solid rock geology, especially where the drift is represented by boulder clay which is generally local in origin, but less so where glacial sands and gravels occur.

In the basaltic area, soils are often reddish, though dark and brown soils are also common. If carefully managed, these soils provide good pasture but have a tendency to fix soluble phosphates, thereby making phosphorous unavailable to plants.

In the Lower Palaeozoic area, soils are very stony but form good agricultural land. In the Schist areas, soils are extremely acid and yield poor

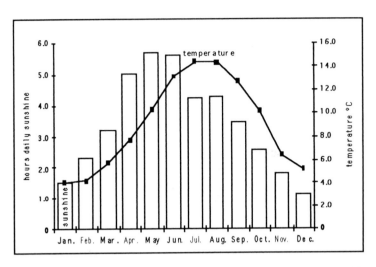

Fig. 1.4 — Average daily sunshine and temperature, 1951-80. No specific weather station was indicated in the original data. However, at this simple introductory level, this is no serious drawback, bearing in mind the comparatively homogenous conditions in Northern Ireland (Source of data: *CSO, Annual Abstract of Statistics, 1989 Edition*).

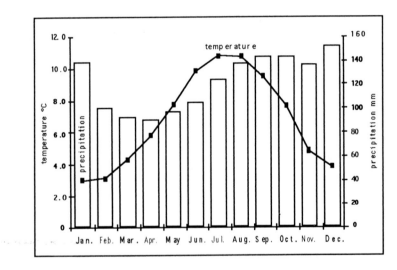

Fig. 1.5. — Average temperature (1951-80) and precipitations (1941-70). Again, no specific weather station was indicated in the data, and this is more serious, since precipitations do vary considerably throughout Northern Ireland. For the distribution pattern of precipitation, see the following Figure (Source of data: *CSO, Annual Abstract of Statistics, 1989 Edition*).

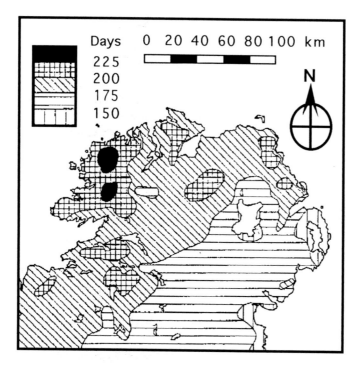

Fig. 1.6 — Average number of days with one or more millimetres of precipitations, 1941-60 (Source: *Atlas of Ireland*).

returns. The Old Red Sandstone in Co. Tyrone forms heavy waterlogged soils, exceedingly clayey and poor in plant nutrients.

Soils on Carboniferous Limestone are in general good as the permeable mother rock provides good drainage, but those on impermeable Carboniferous rocks such as shales and sandstones, become easily waterlogged and yield poor returns.

Areas underlain by Triassic sandstones in the Lagan Valley and south-west of Lough Neagh, as well as on the west side of the basalt scarp in Derry form the best arable land, while the marls give a heavy soil rather suitable for pasture.

Ulster's comparatively cool and damp climate brings about intensive leaching of lime and magnesia from the soils and the production of waterlogged, acidic soils and peats. Only in eastern Ulster is there much arable farming. Steady human intervention is needed to improve conditions for crops (cereals, potatoes, horticultural products). Flax cultivation, once the mainstay of the linen industry, has petered out. Mixed farming predominates, while pasture — already widespread — is spreading further, as many areas in Ulster, especially in the west, are increasingly regarded as only fit for pasture, peat cutting and reforestation.

Man's presence in Ireland is documented from the Mesolithic age. He began, of course, to introduce innovations (new ideas, new technologies) from the earliest times. Mesolithic flint implements were found in Toome Bay (earliest known site, about 6000 BC), Larne, Cushendun and the Bann Valley. Quite naturally, population density was extremely low, technology was simple and the human impact on the environment was accordingly, extremely limited.

With Neolithic man, an epochal innovation — agriculture — was introduced. He left traces of his activity around the coasts of Antrim and Down. Of peculiar interest are the axe factories at Tievebulliagh, near Cushendall, Co. Antrim, and Brockley on Rathlin Island, whose products have been found all over the British Isles. The heavy neolithic axe and shifting cultivation, probably associated with fires, produced some clearings in the forests. These were dominated by oak, with widespread alder and declining elm. Such clearings, however, were sufficiently rare and sparse to allow the forest to regenerate, as indicated by pollen spectra in the bogs.

With the Bronze Age, the early metal axes and a probable increase in population and agricultural exploitation brought about more and more extensive forest clearings. As elsewhere in Ireland, late Neolithic and Bronze Age man built megaliths (court cairns, passage graves, and stone circles), though the most impressive examples are to be found somewhat more to the south.

In the succeeding Iron Age, ring forts (fortified farm-dwellings) and crannogs (lake dwellings) point to a settled population, but also to strife and insecurity. This was probably brought about by rising population and increasing pressure on resources as more and more permanent deforestation occurred. Celt culture emerged some time between 500 BC and 500 AD. Towards the end of that period, christianization occurred. Armagh, in Ulster, became — thanks to St. Patrick — the primate episcopal see for the whole of Ireland. Irish Celt Christianity placed particular emphasis upon monastic life. Irish missionaries spread throughout semi-pagan Europe and gave a decisive contribution to the conversion of entire populations. The *scriptoria* and the schools of monasteries founded by Irishmen in their own country and elsewhere in Europe, made a decisive contribution to the preservation and diffusion of ancient literature and learning.

At that stage in history — it can be stated without exaggeration — the Irish saved civilization. What in the rest of Western Europe goes under the name of "the Dark Ages," is known in Ireland as "the Age of the Saints." Physical isolation played a positive role for Ireland in the context of a heavily unsettled Europe, providing a sheltered position at least until the beginning of Viking raids in the late 8th century. Later, more centrally located European countries began their economic and demographic recovery: a process variously thought to begin from the 9th to the 11th century, and far from simultaneously in the different countries. Then Ireland was fatally left behind. Isolation became a source of backwardness and serious danger.

In the later Middle Ages, Celt society was highly conservative, based on clan structure and comparatively static in comparison with other contemporary Western European societies. In Celtic Ulster, as elsewhere in Ireland, virtually no urban life existed, technology was backward, and war was waged untechnically by means of brave deeds and duels between champions, as in Homer's Iliad.

This made the Irish Celts comparatively weak from a military viewpoint, and military weakness has always invited external aggression — especially when coupled with internal strife and lack of external allies, as was the case for Ireland at that time. Not surprisingly, the Irish succumbed to the Norman invaders in the second half of the 12th century. Nevertheless, sudden reconquests of lost territory took place thanks to terrain conditions, as pointed out above, and to the belated adoption of more up-to-date military techniques (armoury, fighting in orderly formation, and so forth).

Successive conquerors destroyed not only the traditional Celtic society, but radically altered the environment, having in view a more intensive exploitation of the agricultural and pastoral resources of the conquered land. Thus deforestation had become wholesale by the 18th century. Only thanks to recent reforestation, some small areas are nowadays covered with coniferous trees: obviously, the species makeup of these newly forested zones have nothing to do with the original vegetation.

Meadows, moors, peatland and marshes cover most of the territory. The landscape is verdant nearly everywhere, except bare rock outcrops in spots particularly exposed to the occasional fury of the westerlies.

The native vegetation in Ulster, as in the rest of Ireland, is far poorer than in Great Britain. With little more than 900 native species of plants, Ireland's flora has only about two thirds of the species in Britain.

There are, however, some typical plants of outstanding interest. In the acid boglands, Common Sundew (*Drosera rotundifolia*) and Bog Asphodel (*Nartecium ossifragum*) are commonly found. Important calcifuge plants are Foxglove (*Digitalis purpurea*), Bell Heather (*Erica cinerea*) and Sheep's Bit

(*Jasione montana*). Calcicole plants (not very widespread in Ulster) are Cowslip (*Primula veris*) and Lesser Water Parsnip (*Berula erecta*). Arctic-alpine plants are relict of the early "post-glacial" tundra-stage, of which Dwarf Willow (*Salix herbacea*) is to be found especially in Donegal but also in Cos. Antrim and Down.

Hiberno-Cantabrian plant species, probably indicating migration of seeds from the Hiberian peninsula are Wild London Pride (*Saxifraga spathularis*) widespread in Western Ireland, but present in Ulster only in Donegal, and Irish Spurge (*Euphorbia hyberna*) present in the south-west of Ireland, whose only sites in Ulster are those on the Donegal coast. Most interesting, are plants with North American affinities which may prove the migration of seeds across the Atlantic: of these Pipewort (*Eriocaulon aquaticum*) is present in Connacht and Munster, while the only Ulster sites are in Donegal.

A similar poverty is typical also of the native fauna. There are few indigenous mammals (27 species besides whales and seals along the coasts); birds are far more numerous (about 380 species); reptiles are only represented by one species (a lizard); amphibians by three species.

Such comparative biogeographical "poverty" is partly due to the rapid rise in sea level which followed the melting of the ice of the last glaciation, whereby Ireland became isolated before many species had time to recolonize it. However, it may also depend upon the narrower range of ecological niches in comparison with the larger island. The insular and peripheral condition of the whole of Ireland appears therefore not only in its human geography, but also when its biogeography is scrutinized.

However, the scenic value of several areas of the region is a considerable asset for tourism and recreation. The Giant's Causeway has been designated a World Heritage Site. Particularly worth remembering are the famous "Green Glens of Antrim." As is Strangford Lough, an inlet of the Irish Sea connected by a narrow channel to the open sea and having a wide tidal range, so that at each tide, large expanses of mud are exposed; they provide a feeding ground for waders and wildfowl, while several small islands are breeding grounds for many species. Two further valuable nature parks offering good opportunities for birdwatching are the Peatlands Country Park and Oxford Island on the southern shore of Lough Neagh, where the Otter (*Lutra lutra*) can also be seen. Killard, east of Downpatrick, is a coastal reserve with a wide range of habitats. Mudflats and dune habitats can be seen by the mouth of Lough Foyle at Magilligan Point. Rathlin Island off the north-east tip of Ireland has interesting colonies of seabirds. Some of these nature reserves are hosts to the rare protected seabird Roseate Tern (*Sterna dougallii*).

Among the biogeographic changes wrought by man, the most conspicuous are deforestation and diffusion of agriculture, and animal husbandry with cattle, horses, sheep, pigs and poultry. The extinction of the boar took place in the twelfth century and that of the wolf in the eighteenth. Alien species were introduced, such as the rabbit, most probably by the Anglo-Normans, the deer, perhaps in the sixteenth century, and the squirrel in 1911.

Man's impact has thus been decisive, in the north of Ireland as elsewhere, in the shaping of the territory, i.e. the *material space*. More subtly, he has also created a *non-material space* made up by power of territorial control, circulation of ideas, information, innovations. Man has thus introduced *development*. This is understood as *a process of innovative structural change of the society and the space occupied and used by that society*. Innovations are not necessarily positive, they may also be destructive or disintegrative; so that development can accordingly spell either improvement or disaster.

Development, as John Friedmann aptly points out, must not by any means be confused with growth which is a mere quantitative phenomenon. Moreover, development is not merely an economic process, but, more generally, it is also a social, cultural, and political fact. It cannot be understood without account being taken of the physical variables.

Some of the patterns and processes of development can be summarized in a sketch based on the so-called *systems analysis* (Fig. 1.7). Each small rectangle with a tag purports to represent a hypothesized *variable* linked to other variables by arrows indicating probable cause and effect linkages. A multiplier is a set of variables bound to each other in such a fashion as to increase and "multiply" a given process: thus *development multiplier* means the complex of interrelated variables whose effects are "multiplied" and expanded by their interrelations.

For instance, climate, and geological structure and processes affect soils, which in turn affect agriculture and animal husbandry. Likewise, climate and landforms affect siting choice, i.e. the decision to locate a dwelling, lay the foundations of a town, or initiate any kind of economic activity. No human decision, however, is ever taken with a perfect knowledge of the situation; it is rather a result of the *perception* of that situation. A *perception threshold* must be passed before a decision can be arrived at. Several perception thresholds of this kind are indicated in the sketch.

So agriculture and animal husbandry, industry, transport, and services flourish, favoured by the flows of biological and mineral raw materials, while under the decisive impulse of science and technology, population grows.

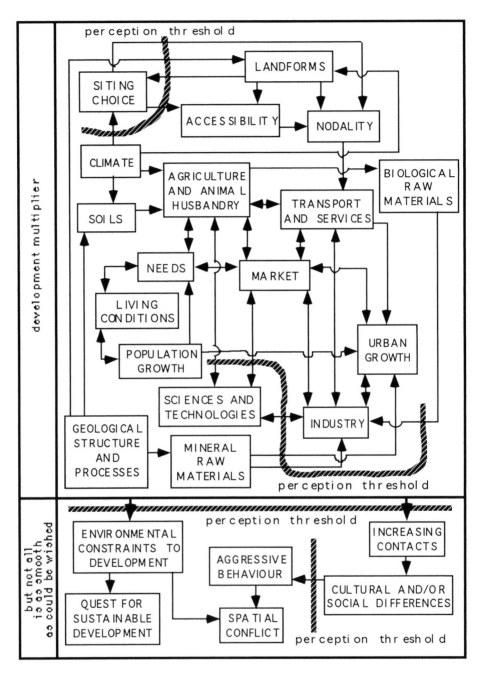

Fig. 1.7 — A sketch of the development process, which promises well, at least initially. For explanation, see text.

This fact, related to the living conditions and needs, fuels the operation of the market.

Everything is well, then? To a large extent, expectations of a better way of life have been fulfilled, at least in the developed world, and therefore also in Northern Ireland, which is a part, though economically depressed, of that world. But, helas, not all is as smooth as could be wished.

CHAPTER 2

CONQUEST, PLANTATION AND LANDLORDISM

After the, initially violent, later somewhat more peaceful, contacts with the Vikings, who left important traces in the rest of Ireland but did not settle in Ulster, Ireland suffered invasions by the Anglo-Normans from 1169.

The Anglo-Norman intrusion into Gaelic Ulster began in the late twelfth century. Their conquest, however, was very incomplete and, in time, they gradually lost ground. The armed forces of the Gaelic chieftains, who were used to going into battle wearing only a light shirt and, as mentioned in the previous chapter, without any strategic or tactical art, in time learnt to fight in a more technical fashion, covered with a coat of mail, and thus became more than a match for the Normans. Many other parts of Ireland, likewise, managed to rid themselves of the invaders.

English pressure, however, became more and more severe under the Tudors, while the Reformation made assimilation between the conqueror and the conquered far more difficult. The change in religion gave birth to a serious cleavage between the English and the Irish, or rather between the Celt and "Old-English" Catholics on the one hand, and the "New-English" Protestant invaders on the other. A strong new feeling of reciprocally hostile self-identity thus arose on both sides.

Ulster was the part of Ireland where Gaelic resistance had up to then proved more successful. While in the rest of Ireland the clans had been forced into submission, the northern Province still largely retained its traditional features. Towards the end of the Elizabethan age, while the rest of Ireland was under colonial control and undergoing a slow process of anglicisation, Ulster, "the great Irishry," where clan chieftains still ruled, was the least anglicised of the four historical provinces. There was practically no town life: the few centres were located on the coast, such as Carrickfergus (Photo 3), and their foundation dated back to the early wave of Anglo-Norman conquest.

Photo 3 — Carrickfergus Castle (late 12th - 13th century). A control point at a gateway of colonial penetration. In the early 17th century it became one of the main power bases of the Chichesters and a grim jail for Sir Con O'Neil of Clandeboy, whose lands were coveted by the invaders (Department of the Environment for Northern Ireland. Crown Copyright).

Elizabeth wished to plant Ulster, but at no expense to the Exchequer. She therefore sought to bring about plantation through private undertakings. Some such attempts were carried out from the 1570s, but were met with little success. Even earlier, however, eastern Ulster had received a persistent trickle of immigration from Lowland Scotland.

The most powerful chieftain, Hugh O'Neill, known as "the Great O'Neill," raised the standard of rebellion against English encroachment in 1594. It was the beginning of the Nine Years War. Fighting on difficult, woody and marshy terrain, which they knew well, the Irish were able to keep the English at bay. In 1598 O'Neill achieved a major victory at Yellow Ford. The news unleashed a new uprising in southern Ireland, where the landing of a small Spanish force, doomed to defeat, took place in 1601. Hugh O'Neill, who had been made *ard rì* (Supreme King), hoped to free all Ireland. He thus gave up the tactical advantage he enjoyed in Ulster and sought to wage war in the South. That was to be his undoing. He was defeated and compelled to submission in 1603.

That very same year, a few days before O'Neill's surrender, Queen Elizabeth died. The new King, James I Stuart, already had great experience in dealing with Celt tribal chiefs: as King of Scotland under the name of James VI, he had been able to weaken the Highland lairds. He set out to apply to Ireland the policy that had proved successful in a similar social setting. He forced O'Neill and the other rebel chieftains to accept the regranting of their lands in full ownership, keeping their vassal chieftains as leaseholders. In this way the communalism-based tribal structure was slowly undermined.

These rather moderate measures were a sore disappointment for the English generals who had hoped for a rich plunder. They had to find other means to carry out extensive territorial usurpation. In the event, not only the Irish chieftains, but the British Crown itself, suffered from it as is apparent from the methods whereby huge landed estates were established, increased, and preserved by the newcomers.

Foremost among them was Sir Arthur Chichester, who earlier had shown his mettle by robbing a tax collector. To escape justice, he took refuge in France, where he earned a living as a mercenary soldier. In 1599, pardoned by Queen Elizabeth, who needed precisely hard and merciless soldiers of this kind in order to defeat the Irish rebels, he took up a command in Ireland. Having proved equal to the task, Chichester rose rapidly within the ruling clique on the island that had been just subdued.

Even before the end of the Nine Years War, the grabbing of the Ulster lands had begun. Irrespective of their loyalty, the Celtic chiefs were dispossessed. Every strategy — whether open resistance, neutrality, or active cooperation — failed to the seemingly insatiable greed of the invaders.

For example, the O'Neills of Clandeboy had been loyal to the Crown, but their strategically located possessions were raided by those whom they had served. Sir Con O'Neill's downfall came about because the English garrison of Carrickfergus looted wine he had purchased and imported from overseas. His servants reacted and in the ensuing affray one of the soldiers was killed. The scuffle was dubbed "treason" by Sir Arthur, who had Sir Con apprehended, jailed in Carrickfergus Castle and tried as a rebel. Before the inevitable execution, Con escaped and reached London with the help of a Scotch laird, Hugh Montgomery, who took him to see a royal favourite to secure a pardon against the forfeiture of half his estate. At court, a former spy pleaded mercy for him in exchange for an additional slice of his estate. In all, Con was thus shorn of two thirds of his possessions, merely because his servants had used force to repeal robbery and violence.

In 1603, thanks to the interest of Lord Devonshire, then Lord Deputy, Chichester obtained a letter from James I granting him the lifetime governorship of Carrickfergus, as well as lands between the castle of Carrickfergus and that of Belfast.

To understand what follows, it is necessary to recall some peculiar legal practices of seventeenth century England. These are well expounded in the seminal studies by Healy, on which a part of this chapter is based. I have also scrutinized the original evidence, preserved in the Public Record Office of Northern Ireland (henceforth shortened as PRONI), in order to ascertain whether Healy's conclusions stood the test of time.

Frauds began after a royal letter, granting lands and privileges, had been issued. Such a letter was a warrant given to an applicant, who had usually submitted it to the King in draft form, written on a parchment. If the sovereign agreed, the letter was sent to the Signet Office to be copied into the Signet Book.

A letter concerning grants in Ireland had to be taken to Dublin, where the Law Officers gave their "fiat" to make it "patent," i.e. public. The patent ought to have been a mere translation in legal form of the King's warrant. Often, however, thanks to conniving officials, one could extend the grant much further than the King had wished. The patent was then sealed under the Great Seal of Ireland by the Lord Chancellor who "usually" ordered it to be copied onto the parchment rolls of the Chancery.

No register of the patents was kept, and the loopholes in such an embrionic administrative system made all sorts of deceptions possible. For instance, if the grantee omitted to have his patent copied onto the Chancery rolls, the Crown was left without a copy and it was impossible to tell, especially after a few years, which lands had been granted.

Sir Arthur Chichester, on the strength of the royal letter, took not one but two patents. In one of them he "annexed," besides the site of Belfast, the whole Lagan river, though the letter only granted him a fishery close to Belfast castle. With the second patent, concerning the governorship of Carrickfergus, he took over, it seems, the whole Lough Neagh. None of the patents was enrolled in the Chancery, in order to ward off a discovery. Both patents, according to Healy, later disappeared.

A new Solicitor-General, Sir John Davies, arrived in Dublin in the month of November 1603 and Chichester very soon made him a confederate in his attempt to get a new royal letter covering the seizure of Lough Neagh. It was clear, in fact, that the patents would not have borne scrutiny in a law court, should Sir Arthur have sought to uphold his claim. Lough Neagh had from time immemorial been *res communis* of the local inhabitants. Even the right of the King to grant it was highly dubious, except by right of conquest.

The new royal letter, issued in 1604, granted to Sir Arthur the command of the patrol boats of the lake, but the derived patent spoke instead of the "fishery of the said Lough up to the Salmon Leap on the river Bann." In the PRONI is preserved a Latin parchment from the 9th of May 1604 (D.509/2), almost certainly the royal letter referred to by Healy.

The document confirms to Arthur Chichester and his heirs the granting from the 5th November 1603 of Belfast castle and the lands named Le Fall (Falls) and Moyellon (Malone), moreover the estate close to Le Therogh otherwise called Le Sinnament (from the Gaelic *tuogh*, territory, and the Anglo-Norman *cinnament,* landed possession), for a breadth of seven miles along the coast from Carrickfergus towards the said castle (of Belfast), and also the waters of the Lagan and the Mavist (?), with all their appurtenances (buildings, meadows, woods, pastures, fisheries, etc.), and moreover confirms a previous grant of the territory of Clandeboy, close to Belfast castle, and of various hamlets in the vicinity, which had belonged to Rudolph Lave (or Law?), lately deceased, to whom they had been granted by Elizabeth I on the 6th June 1598. The document, however, does not mention Lough Neagh.

To the considerable difficulties in reading, due to the precarious state of preservation of the parchment, one must add those encountered in attempting to pinpoint localities with a modicum of precision, since place names were standardised by the Ordnance Survey only from the 1830s. Prior to that time, the name of any place could be written with any number of different spellings. In the case of this document, the identification of at least some localities has been made possible by means of comparison with documents of the inquisitions of the Chancery of Ireland, carried out between 1585 and 1618, published in 1829 under the title *Inquisitionum Cancellariæ Hiberniæ Repertorium* and preserved at the Ulster Folk and Transport Museum in Cultra, Co. Down.

On the 15th of October 1604, Chichester was appointed Lord Deputy, and his newly acquired power paved the way for renewed attempts to secure Lough Neagh. The witchhunt atmosphere at the time of the alleged "Gunpowder Plot" (5th November 1605) made any encroachment upon Catholic rights far easier than usual. Takeovers were usually perpetrated through accomplices. Sir Arthur found a confederate in the Scottish adventurer James Hamilton, to whom he granted, by means of a patent, the fisheries of Lough Neagh and those of the Bann (14th February 1606). Immediately afterwards, he got a reconveyance by Hamilton of the very same properties. The whole business was kept secret.

The practice of secret property transfers for the purpose of fabricating evidence of legal titles to estates had been outlawed in England by Henry VIII. Only in 1634, when Chichester's plots had come partly to light, the same prohibition was extended to Ireland.

Another trick was that of reviving an old dispute between Hugh O'Neill — whose huge estate included the present counties of Derry and Tyrone, as well as large parts of those of Armagh and Monaghan — and Sir Donal O'Cahan, one of his tributary chieftains, whose clan occupied the territory of present-day Co. Derry. During the war, the English generals, Sir Arthur Chichester and Sir Henry Dowcra, had promised O'Cahan to free him from tribute to Hugh O'Neill if he betrayed his liege lord. However, in the last days of Queen Elizabeth, knowing that a Scottish King was likely to ascend the English throne, and fearing the newcomer would be better disposed towards the Irish, Chichester had promised Hugh O'Neill total pardon including the right to keep his lands entirely, with the attendant tributes of the chieftains subject to him.

It was obviously impossible to keep both promises. The dilemma was brilliantly solved by keeping none. Initially, however, the one made to O'Neill was kept. In 1606, Chichester began to encourage Hugh Montgomery to support O'Cahan, who thus stopped paying tribute to Hugh O'Neill, in violation of the Mellifont treatise, which had put on end to the war and restored to O'Neill all his former rights.

Hugh O'Neill sougth to enforce payment according to the law. He seized O'Cahan's cattle and the Bann fisheries. It was just what Chichester and his confederates were waiting for. Sir Arthur started a lawsuit without informing the defendants of the fact that he himself possessed spurious patents for the fisheries. John Davies, in his capacity as Solicitor-General, wrote to Cecil Lord Salisbury, at the time the most influential of the ministers, stating that Hugh O'Neill's patent was defective. He did not say, however, that a petition by O'Neill to have the same patent mended had been rejected upon advice from Davies himself; and this at a time when "defective" property titles to Irish lands were mended for the asking when they belonged to English or Scottish adventurers, unless utterly indefensible like those of Chichester.

The verdict, therefore, went against Hugh O'Neill, whose patent was declared not valid. All his lands were forfeit to the Crown, and the luckless man was bound go to London and put himself at the mercy of the King. The Earl evidently came to the conclusion that he was a doomed man whatever he did, either disobey the order and stay in Ireland or go to London, where anti-Catholic sentiments, fired by the fresh memory of the "Gunpowder Plot" still ran high. He risked ending up in the Tower, where so many Irish noblemen had already died. Thus the Great O'Neill, the man who for nine years had fought the English in the field, had not enough courage to withstand further legal persecution and decided to flee the country, together with many other harassed chieftains. This was the well known "flight of the Earls." They sailed in 1607 from Rathmullen to seek refuge in France. Such was the terror spread by Chichester that their ship was denied water supplies in Donegal.

Irish nationalists have deplored this flight, which left the field entirely to the foreign adventurers; but it is extremely unlikely that O'Neill might have gained anything at all by sticking to his guns, except putting his own personal safety in jeopardy. The power of the newcomers, both in the military forces at their disposal and in their influence at Court in London, was overwhelming. They proceeded inexorably according to the logic of a closed élite, so often successfully employed against weaker opponents in many other lands and historical situations, in three classical stages: (i) achievement of a monopoly of power, thereby laying a control surface upon the whole territory; (ii) ruthless plunder of the victims, thereby achieving an economic monopoly; (iii) crushing of any resistance.

The lands of the Irish clans were thus easily seized. O'Cahan was the first of the chieftains who had stayed behind, to suffer the ruin of the traditional power structure he had helped to come about. Apprehended on a flimsy pretext, he was at first jailed in Dublin Castle, then in the Tower of London, where he died twenty years later. A similar treatment was meted out to all other cheftains. The last one was a minor of the O'Reilly clan in Co. Cavan; there was no difficulty in taking his lands.

Chichester gave the fishery of the Bann to his nephew Arthur Basset (PRONI D.389/1), while the legitimate owner, Sir Randal Mac Donnell, appealed in vain to the government in London. Within six months Basset regranted the property to his uncle and was rewarded with a knighthood.

Like all colonies, Ireland was at the same time under two élites: the metropolitan one, in this case the Crown and the Court in far away London, and the élite locally settled. While the latter was bent on grabbing lands, the metropolitan élite — unlike in other colonial settings where some kind of protection for the natives was not entirely forgotten — was by no means interested in protecting the Irish. In spite of that, the vested interests of the two power groups were not the same. When the King was informed that Ulster had been deprived of its indigenous élite, he decided to undertake a plantation.

Unlike former Irish plantations, that of Ulster was carefully planned on purpose-made maps which can be found today in the British Museum (B.M.Cott.Aug. 1/11/44) and in the PRONI (T.2528/7A and T.535). The introduction of modern cartography in Ireland occurred just to serve the needs of the plantation.

The six escheated counties (Armagh, Cavan, Derry formerly named Coleraine, Donegal, Fermanagh, Tyrone) do not coincide entirely with the six which currently make up Northern Ireland (Antrim, Armagh, Derry, Down, Fermanagh, Tyrone).

In 1610, the English government laid down the conditions for the plantation. The lands seized from the Catholic Church were allotted to the Anglican Church. Derry County was enlarged at the expense of Tyrone and granted to twelve different corporations of the City of London (hence the official name "Londonderry"); its territory was divided into large estates of 3000 acres each.

Lands that had belonged to the clans in the other five counties were divided into baronies, each of which was to be given to grantees of one type only. There were four types: (i) English undertakers; (ii) Scottish undertakers; (iii) English servitors, i.e. men who had served the Crown, usually as soldiers; and finally (iv) Irish natives, presumably chosen for their collaborationistic propensities and docile disposition. Servitors and natives were allowed to hire local manpower. Undertakers were required to bring from England or from Lowland Scotland twenty-four able-bodied men for every 1000 acres of fertile land before November 1611. A quarter of the land went to the Anglican Church. Other grantees included Trinity College, founded in Dublin by Queen Elizabeth in 1592, as well as towns, fortresses and schools.

A network of towns was set up as a tool of territorial control, as well as to provide market, administrative and service centres. These towns had a regular gridiron layout, with a central square called "the Diamond." The consciously imitated model, under both functional and structural viewpoints, was the Roman one. The municipalities were run by corporations set up by royal patent. Their members, usually twelve, sat for life. When one of them died, the others coopted a replacement. This was obviously the ideal recipe for a closed, immobile gerontocracy. The elitarian closure was strenghtened by the exclusion of Catholics from the freedom of the town, in which all citizens had to take the Oath of Supremacy. Catholic inhabitants were confined in a separate "Irish town," outside the walls, as in the rest of Ireland.

The towns were also important seats of political power. Each of them had the right to send two representatives to the Irish Parliament. According to the original plan, each town was allotted land which was expected to provide an early start for economic development. Serious difficulties, however, were encountered: most towns became rotten boroughs at the mercy of the landowners.

The Lord Deputy Sir Arthur Chichester suggested in 1610 that an undertaker or a servitor close to each one of the projected towns, should be made superintendent over it and take charge of the building of houses and the recruitment of inhabitants. As a reward, he would get the lands originally allocated to the town. The suggestion was accepted. This led inevitably to a land grabbing rush which enriched the adventurers further, and dramatically depleted the income of the infant municipalities.

The new settlement pattern followed closely the traditional one. This was based upon a minimum territorial unit, the *baile* (English townland), which in Ulster was called *ballyboe*, and, on an area of 60 acres, housed a farm or a hamlet. A group of 16 ballyboes made up a *ballybetagh*, a territorial unit subject to a tribal chieftain. Actually, ballyboes and ballybetaghs were far more extensive, five times the nominal area or even more, since only the productive area was measured, in terms of heads of cattle the land could support. It must not be forgotten that the whole traditional economy and society were centred upon the cattle. Cattle raids were regarded as deeds of bravery, and the greatest Irish epic poem, the "Táin Bó Cuailnge," on the hero Cu Chulainn, deals precisely with a war between *Uladh* (Ulster) and *Connachta* (Connacht) for the possession of a sacred bull. On the other hand, the Irish did not measure actual areas at all: they were not interested in the problem and had neither conceptual nor technical tools suited for the task.

The various ballybetaghs, though of different size, were approximately equivalent in productivity. The settlers usually received whole ballybetaghs, including about 1000 acres of fertile land and far larger stretches of unproductive or nearly unproductive land. The main undertaker of each barony was granted three ballybetaghs, the others one or two. Every undertaker was under the obligation to build a fortified farm (bawn) and, if he had more than 1000 acres, a brick or stone house. The territorial organisation had been similar under the Anglo-Normans (who, in turn, had taken over the traditional land subdivisions of the Gaelic clans), so that, as a whole, a decided continuity can be detected in the fabric of the countryside.

The most significant innovative impact of the plantation was felt in the economic domain, through the introduction of the market economy, which replaced the traditional subsistence economy. The urban network, the building techniques based upon fortified stone or brick houses and farms or upon timber houses different from the Irish twin houses, the new landownership system replacing the communal property of the clans, were the other major innovations in the field of material culture.

A key innovative event was that nearly all the settlers were English speakers with the exception of some Scotsmen from Galloway who spoke Gaelic as a first language. The other basic innovation and an endless source of social cleavage was the Protestant religion of nearly all the settlers, with the exception of a few Catholic groups, also Scotsmen, followers of Catholic undertakers. One such group settled in the barony of Strabane, and this accounts for the persistent overwhelming Catholic dominance in the area. Pretty soon these groups suffered discrimination like the natives. The main social barrier appears to have been, from the very beginning, the confessional one: at an earlier stage the natives were no longer called Irish but, with contempt, "papists".

However, Catholicism could not be uprooted. The Gaelic heritage also persisted. Place names remained mostly Gaelic: Belfast, for example, stems from *Beal-Fersat*, i.e. "the ford of the sand bank." English is spoken in Northern Ireland with a peculiar accent and intonation and many local variants, all of which reveal a resilient underlying Gaelic substratum of both local and Scottish origin.

The plantation proceeded sluggishly, due also to the lack of enthusiasm of Sir Arthur Chichester and his friends, who obviously preferred to deal with defenceless natives rather than with settlers under the King's protection. Only around 1622, the planned number of planters was reached (Fig. 2.1). Many undertakers had not been able to attract a sufficient number of Englishmen or Scotsmen and, therefore, employed Irish hands in open

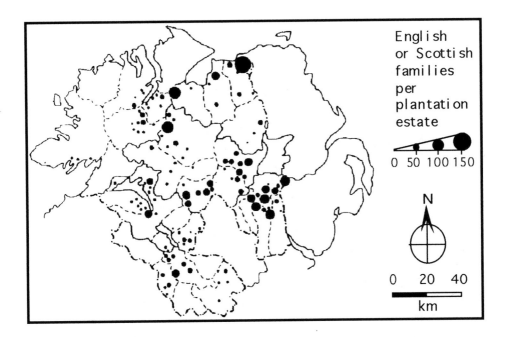

Fig. 2.1 — English and Scottish settlers on plantation estates in 1622. (Source: Robinson).

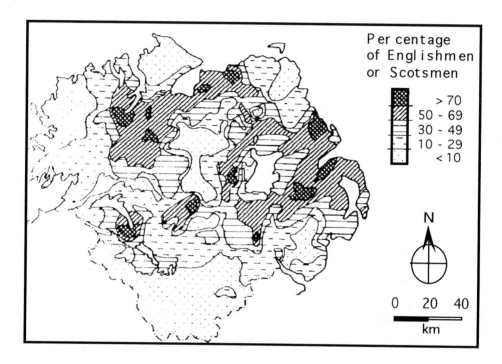

Fig. 2.2 — English and Scottish settlers in percentage of total population by parish, around 1659. (Source: Robinson).

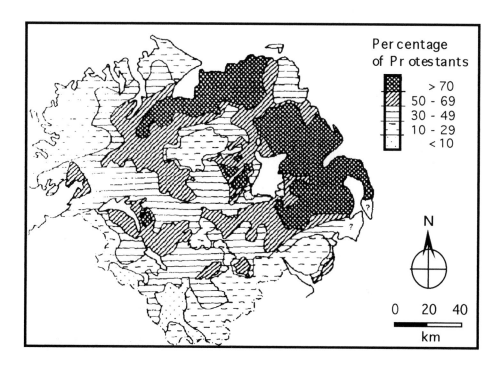

Fig. 2.3 — Protestant inhabitants in percentage of total population by parish, around 1766. (Source: Robinson).

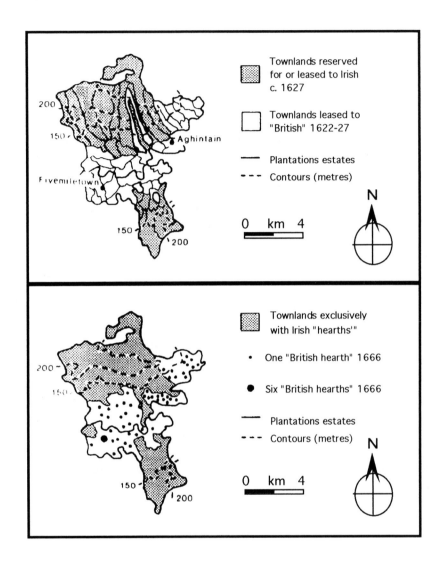

Fig. 2.4 — Ethnic segregation between the "British" and the Irish on two plantation estates in the Clogher Valley, Co. Tyrone, in 1627 and 1666. (Source: Robinson). Strictly speaking, the term "British" is inappropriate, since Great Britain came into existence only with the Union of England and Scotland in 1707. By "British," in this case, is meant either an English or a Scottish settler.

defiance to the conditions they had subscribed. By the article 26 of the "Graces" of 1628, the King allowed these undertakers to keep their local manpower, provided they paid higher taxes and contained the Irish within only a quarter of their allotted lands.

Only by gradual immigration, notably from Lowland Scotland, did the Protestants achieve the numerical majority in most of Ulster (Figs. 2.2 and 2.3).

According to a taxation register from Co. Tyrone dating back to 1660 (PRONI T.1365/3), studied by Robinson, 32% of the men with English or Scottish surnames are registered as farmers and 68% as labourers or servants, while the percentages for Irish surnames are, respectively, 28% and 72%. This difference seems to indicate the absence of a marked social imbalance between the two groups, but one has to bear in mind firstly, that before the conquest, the land belonged only to Irishmen, and moreover that Englishmen and Scotsmen usually received the best lands, while the natives were granted the worst ones, as shown, for example, by the situation in two estates in the valley of the Clogher, Co. Tyrone, studied by Robinson (Fig. 2.4). The statement by Moody that the expulsion of the Irish from the more fertile lands and their replacement by planters was not brought about by violent usurpation, but was only the result of a gradual operation of "economic forces," sounds awkward indeed.

The inflow of settlers was a substantial danger for Chichester's ambitions. When Sir Arthur was informed that the grant of Co. Derry to the London corporations was to include the Bann and Lough Foyle, he took active countermeasures. Acting as a judge and heading a jury which included Davies and other confederates, he held an inquest on the properties in Ulster. Thus he could pass sentence in 1609, at Limavady, to the effect that the whole Bann, including the bed and the soil, from the mouth at Coleraine to the source in Lough Neagh, belonged to himself.

Early in 1610, the City of London accepted the grant of Co. Derry (Londonderry, if you must), including the Bann and Lough Foyle. When James I was informed by Mac Donnell, the legitimate owner of the fisheries of the Bann, that three quarters of the tidal river had been granted to Hamilton, he did not realise that Chichester had seized all the rest of the river and Lough Neagh. The King then promised to purchase back what he thought had been "inadvertently" granted to Hamilton. The latter submitted an obstreperous bill, claiming he had paid, together with seven (phantasmagoric) associates, 4,760 pounds to purchase the properties of "various persons" in the Bann and Lough Foyle, plus legal expenses. This account was corroborated by Chichester in his capacity as Lord Deputy, and the King paid Hamilton the handsome sum of 2,500 pounds, while 2,000 pounds were disbursed by the City of London.

Sir Arthur then hatched a new plot, issuing a deed of conveyance in which he surrendered to the King all his pretences to the Bann and Lough Foyle. Witness to this deed was the Anglican bishop of Dublin, Jones, who being Lord Chancellor of the Kingdom of Ireland was empowered to accept gifts to the sovereign without the formality of a royal letter. The King, on the other hand, was never informed of the "gift," which, according to Healy, was only meant to provide evidence of Chichester's right to the fisheries. It could also prove what a "disinterested" man he was, just in case anyone discovered that the industrious Lord Deputy had grabbed 10,000 pounds belonging to the Crown out of the income of the lands confiscated in Ulster.

In 1613, a Parliament was set up in Ireland, obviously entirely controlled by the conquerors. The first piece of legislation of the new body was to proclaim O'Neill and the other Ulster chieftains under attainder for high treason. This gave official sanction to the forfeiture of their lands, thereby strenghtening the position of Chichester and his associates.

A couple of years later, however, Sir Arthur fell in disgrace. Perhaps King James had received intelligence about the 10,000 pounds Chichester had pocketed, or else he was angry at the further 10,000 pounds obtained from the Irish Parliament by the Lord Deputy "for extraordinary expenses," or he had not yet "digested" the 2500 pounds he had been cajoled into giving out for the Bann and Lough Foyle, or maybe he was beginning to harbour suspicions about the land grabbing activities of Chichester, whose powerful protector at Court, Cecil, had died in the meantime.

The first hint of a change was a letter dated 25th of March 1615, in which the King expressed dissatisfaction with the little zeal whereby the settlement of colonists was being supported. With another letter dated 22nd of November the same year, James removed Sir Arthur from the post of Lord Deputy. The Irish Parliament had been dissolved before it had time to pass legislation supporting Chichester's faked property titles. According to Healy, the fall of Sir Arthur was brought about mainly by the grievances of the City of London for his attempts to seize the Bann and Lough Foyle fisheries.

In spite of this, Chichester's fortunes did not collapse thanks to his associates who still held key posts in Ireland's colonial administration. Already the following year, Davies e Oliver St. John (who had replaced Sir Arthur as Lord Deputy), managed to get for him the post of Lord High Treasurer of Ireland. In his new capacity, Chichester had under his control the rent rolls, and could thus conceal, for the time being, his abstraction of the revenues of the forfeited lands.

Sir James Balfour, a Scotsman who earned a living as a professional informer on the malpractices of estate magnates against the Crown indifferently for a reward from the King, or from the landowners he used to blackmail, began to investigate Chichester's acquisitions. As a counter-measure, Oliver St. John sent to London the vice-treasurer Sir Francis Blundell who, thanks to the support of the royal favourite, the marquis of Buckingham, managed to put an end to the inquest. Balfour was silenced by means of grants of Ulster lands.

Even if the investigation had been carried on, it would not have been easy to clear the matter up: Sir Arthur had a second line of defence ready thanks to the control he exercised, in his capacity as a treasurer, on the rent rolls of the Exchequer on which he had inscribed a debt of his of 12 shillings and 6 pence as rent for the Lough Neagh and Bann fisheries. The almost casual entry is dated 1618, precisely the year of Balfour's inquest, and is tantamount to a silent acknowledgement of Sir Arthur "rights" from the King. No similar entry appears for the previous or the following years.

After Balfour's threat, however, Chichester sought to get a wholesale grant to secure his booty once and for all. So, in 1619, with Buckingham's support, he made efforts to get a new royal letter. The following year the King agreed, but, according to Healy, only to confirm the previous grants.

A printed transcription of the patent taken on the basis of this letter exists. The document, dated 20th November 1620 and signed by John Lodge, Deputy Clerk and Master of the Rolls (PRONI T.712/26), concerns lands in Co. Antrim, besides the fisheries of Lough Neagh and the Bann. It contains a *non obstante* clause, i.e. a clause which declared it valid notwithstanding that the prescribed inquest had not taken place. The inquest was mandatory as it was necessary to ascertain whether rights of third parties existed upon the properties granted by the King. Thus the clause was an obvious illegality. However, very soon the King ordered the lawful investigation to be held.

Chichester easily overcame this difficulty by packing the commission with his friends. The principal commissioner was Stephen Allen, the Escheator for Ulster, who owed his post to Chichester. In a seating at Derry, controlled by the London corporations, the commission ruled that the Bann belonged to those corporations, free from any burden. A fortnight later, at Carrickfergus, the main centre of Chichester's power, the same commission ruled that the river belonged to him. Only the latter ruling was taken into account by the colonial government in Dublin, so that Chichester was given the 25th of November 1621, a patent in which both the Bann and Lough Neagh were declared his property.

From this wrangle between a cunning and grabbing peripheral power (Chichester's colonial "mafia") and a corrupted and absent-minded central power, the Irish were utterly absent. Defeated and deprived of their indigenous élites, they did not even manage to obtain information on the struggle raging above their heads, which was of no consequence to them anyway: no matter who won they were destined to servitude. The spatial system had significantly changed: theatre of the struggle and decisional centres were the cities — Derry, Carrickfergus, Dublin, and far away London — alien and hostile to the Irish.

The next phase of the contest was acted in London, the centre of supreme imperial power. The merchants of the City enjoyed the geographical advantage of being close to the King, unlike Chichester, who had to operate at a distance, relying on royal favourites, and receiving information on what took place at Court too late, given the slow communication infrastructures of the age. It is very likely that the corporations of the City, informed of the deception they had suffered from the commission of inquest, hastened to complain to the King.

All of a sudden, in January 1622, Chichester was sent on a diplomatic mission to the Palatinate, and at the same time his accomplice Oliver St. John — in high glory up to that moment, so much so that he had just been made a Peer of England — was removed without warning from the post of Lord Deputy. Both were ordered to leave Ireland immediately, though in a solemn and honourable fashion to avoid staining the imperial image in front of the natives. In London, however, the King publicly reproached Chichester for his malpractices.

When the the odd couple had departed, James I issued, in May 1623, strict orders to keep all land grants and all confirmations of previous grants, until a secure method had been found to protect the properties and revenues of the Crown in Ireland. Moreover, Chichester was compelled to lease in perpetuity, the Lough Neagh to the London corporations for 100 pounds a year. Sir Arthur died in London on the 25th of January 1625. Three months later, King James had also died, and the new King, Charles I, took steps to compel Sir Arthur's son and heir, Edward to make good the 10,000 pounds embezzled by his father from the revenues of the lands of the forfeited estates.

According to Healy, Lord Strafford, made Lord Deputy of Ireland in 1633, declared that very year that in all the Plantations, the Crown had sustained shameful injury, by passing in truth ten times the quantity of land expressed in their patents.

In 1635, the Court of the Star Chamber declared forfeit the grant of the Ulster lands to the London corporations. The latter petitioned King Charles to be freed from the obligation of paying the Chichester family the lease of 100 pounds a year for Lough Neagh, which had become useless to them, since they had to give up the Bann too.

Lord Strafford then ordered an inquest on the property titles of the Chichesters. Sir Arthur's illegalities were then partially uncovered, and Strafford demanded that the family gave up Lough Neagh. The Chichesters accepted, but applied for an yearly "compensation" of 60 pounds and an unimpeachable patent for all other properties. Later, in 1639, having realized the plots of the Chichesters for securing the Bann, Strafford demanded that they give up any claims to it as well. The new patent was issued in 1640, but the Chichesters failed to enroll it in Chancery and kept its stipulations secret, hoping to recover Lough Neagh and the Bann in "better" times. Nor was such hope entirely ill-grounded, though they had given up both the lake and the river in writing. Everybody knew that a renunciation was quite often the prelude to a regrant of the same property to the previous owner.

Then followed the stormy times between the outbreak of the Irish uprising (1641) and the Restauration (1660). The Chichesters and their allies survived unscathed, they did not suffer in any way for being, in all probability, one of the causes of the insurrection. In 1641 the garrison of Carrickfergus, commanded by a Sir Arthur (son of Edward, and therefore a nephew of his namesake, founder of the dinasty), in one night, slaughtered all Catholic tenants, men, women and children in the peninsula of Island Magee, who had shown no inclination whatsoever to take part in the incipient uprising.

The plantation was later wiped out by the insurgents, and only after the bloody reconquest by Cromwell (1649-1650), could it be re-established. The new Sir Arthur, however, stood secure in his stronghold at Carrickfergus which the rebels were unable to conquer. There is no evidence at all that he punished or even reproached anyone for the atrocities at Island Magee; nor did he justify or explain in any way what he and his men had done. On the other hand, no one called him to task, as the only people who might wish to have justice, the Irish insurgents, could not lay hands on him.

The Chichesters continued to hatch schemes for grabbing the fisheries under Cromwell's republic, usually through some confederate. In 1654, the

City petitioned Oliver Cromwell to get Co. Derry back. The Lord Protector agreed, and granted the Londoners, by means of a patent registered at Westminster and Dublin in 1657, all that had been taken away from them by Charles I, including the fisheries.

In spite of that, the most powerful municipality in the British Isles was victim of a new plot by Sir John Clotworthy, a typical character of the Anglo-Irish Protestant colonial society of the time, who became known for his axiom: "Religion should be preached in Ireland with the sword in one hand and the Bible in the other." Appealing to Cromwell, Clotworthy got a lease of the Lough Neagh for ninety-nine years as a reward for service. Back in Ireland, he had the grant illegaly extended to the Bann, thanks to the connivance of Oliver's son, Henry Cromwell, who ruled Ireland. When the grant was recorded, Clotworthy applied to Henry in order to get, through John Thurloe, secretary to the Lord Protector, a new Signet Letter covering the fraud. Thurloe, however, ignored the request. Two grants of the Bann, mutually incompatible, thus came into being.

In July 1660, two months after the Restauration, Sir Arthur Chichester, just made Earl with the title of Lord Donegall for services rendered in Ulster, set out on a journey to London to pay homage to Charles II and to recover the fisheries. A serious difficulty, however, was the surrender of Lough Neagh and the Bann subscribed in 1640. To accuse Strafford of an "injustice" was inconceivable in the changed political situation, as he had perished on the scaffold for his unshakeable allegiance to Charles I. Chichester, therefore, put the "blame" on Wandesforde, a long dead subordinate of Strafford who had merely obeyed orders.

Late in 1660, Clotworthy, now made Lord Massereene, obtained confirmation for his lease of the fisheries. He was interested in the success of Lord Donegall's application for a royal patent including Lough Neagh and the Bann, because the two together would have been in a stronger position to withstand the pressures of the City. On the 28th of February 1661, a letter from Charles II transferred to Lord Donegall, besides the fisheries, also the 40 pounds lease paid yearly by Lord Massereene. Instead, on the 10th of April 1662, the new grant of Co. Derry to the City was sanctioned; once more it included the Lough Neagh and the Bann, as if these had not just been granted to Lord Donegall.

To comprehend how such legal monstrosities could take place, one has to bear in mind the admistrative muddle, the venality, the moral relaxation of the Restauration. No true bureaucracy, capable of monitoring the grants and preventing their duplication, was yet in existence. Taking bribes was a habit;

the King himself and all the great men never refused "presents." Gifts to "deserving" people were openly sanctioned in Parliament.

In June 1661, with his Signet Letter, Lord Donegall betook himself to Dublin, where no Lord Deputy had yet been commissioned, so that three judges were acting in his stead. He applied for a patent and got it by the 3rd of July (see a copy of it dated 2nd of June 1879, PRONI T.712/6), after only ten days, while usually years went by between the submission of a Signet Letter and the issue of the patent. The delay was not due to mere inefficiency; several lengthy formalities had to be attended to, of which the most important was the preliminary public inquest mentioned above, without which the patent could not be valid. The three judges, Lord Chancellor Eustace and councillors Coote e Boyle, instead, inserted in the patent one more *non obstante*: i.e., as Healy aptly remarks, a further illegality, like the one, previously mentioned, of 1620.

The procedure was so risky, even in such a "disinhibited" social environment as far as property titles were concerned, that Lord Donegall applied for a further royal letter to conceal the deception. He got it and on the strength of it, sought to obtain a new patent. Before that was ready, however, the Irish Parliament passed the Act of Explanation of 1665, intended to secure the estates of landowners. The new Act intimated that extant grants would become void unless registered within two years.

A deed of revocation on 13th of April 1665 exists, as two copies (PRONI D.389/7 and D.509/15), by Lord Donegall and his wife, of a previous grant of 28th of June 1656, by the same Lord Donegall, for Lough Neagh and several lands in Cos. Antrim and Down to Sir William Hicks, Sir John Temple and Sir Audley Mervyn. Two days later, Lord Donegall issued a grant for the same lands and fisheries to Sir Chichester Wrey and again to Sir Audley Mervyn. Such documents were kept so secret that over two centuries later, in 1878, in a debate at the House of Lords on the property of the fisheries, Lord Cairns stated that between 1660 and 1811 there was a total void concerning anything that might indicate property or occupation rights. Healy, who carried out the most diligent investigation on the matter, does not quote these documents.

The goal of such a maneuvre (illegal in Ireland from 1634, as mentioned above) was the usual one: to create a precedent showing that Lord Donegall enjoyed the full possession of the fisheries and could dispose of them at his pleasure. The same object was pursued by some later documents of the 8th July 1668 (PRONI D.389/8), 18th July 1670 (PRONI D.509/23), 20th July 1670 (PRONI D.389/9), whereby Lord Donegall granted rights on several possessions, including the Lough Neagh and Bann fisheries, and subsequently took back the same rights again, always with the same people; the name of Sir Audley Mervyn constantly appears, sometimes those of Sir Chichester Wrey, Sir William Hicks (Lord Donegall's father-in-law) and Sir John Temple. The latter was Master of the Rolls in Ireland, and therefore a highly authoritative confederate.

There is, on the other hand, a similar and more ancient document for the same purpose: a deed of reconveyance from 20th June 1625 of lands in Cos. Antrim and Donegal and of the Lough Neagh fisheries to Viscount Edward Chichester, son and heir of Sir Arthur, from Sir Faithfull Fortescue, Arthur Usher, Tristram Beresford and Charles Pointz (PRONI D.509/12). As the father had just died, if had obvioulsy appeared advisable that the heir should be in possession of a deed supporting his property rights.

This device was just an additional line of defense. The chief weapon would have been a new patent which, however, in 1667, five years after the application, had not yet been issued, while the deadline set by the Act of Explanation was ominously approaching,. The first patent had been issued, as mentioned above, after only ten days. What made the Anglo-Irish power group in Dublin so reluctant to satisfy the requests of the family? The heart of the matter was that no helpless Irish chieftains were to be dispossessed, but the powerful London municipality.

So Lord Donegall had just the patent of 1661, illegal, as we have seen, since it was not preceded by the lawful inquest, and that of 1640, issued by Wandesforde, Strafford's secretary, which excluded the fisheries. No other way out was in sight than having the 1640 patent enrolled. The attempt by the Chichesters to grab the fisheries was thus thwarted, but only for the time being. Meanwhile they were left in possession of a huge estate, in which they oppressed and ruthlessly exploited their tenants.

The behaviour of the Chichesters in this regard was by no means isolated. Many of the land grants in the Elizabethan and Stuart ages clearly reveal a design not only to grab land, but to reduce the natives to slavery. For instance, the charter of Queen Elisabeth to the Smiths (1571) granted, besides lands in Co. Antrim, also all the inhabitants, male and female, as chattels. Likewise James I awarded Hamilton (1605) not only land but "native men and women villeins and their followers." In 1613, the Corporations of the City of London were allowed by royal patent to take "estrayed bondmen and bondwomen and their followers."

Discrimination was the rule in Ireland. Outside Ulster, Catholic Anglo-Irish landowners present in large numbers, were deprived of their lands in the same ruthless manner as the Irish chieftains. Non-Anglican Protestants, of whom by far the largest group was that of the Ulster Presbyterians of

Scottish stock, were also the object of severe discrimination, especially in the Penal Age. But beyond official discrimination, abetted by the Anglican power groups, the boundless greed of the landowners had even more serious consequences. Together with the short-sighted and egoist policy of the British government and the sickly stubborness of King George III, Anglo-Irish landowners contributed substantially to the loss of the North American colonies.

The fifth Earl and first Marquess of Donegall, already made enormously rich by the growth of Belfast — which was so entirely in the hands of the Chichesters that the town did not even vote for Parliament, but its two representatives were appointed by Lord Donegall — and further enriched by the rents paid by his tenants, nevertheless was unsucceessful in covering the huge expenses of a profligate lifestyle. When a great many leases in his Co. Antrim estates expired at the same time, Lord Donegall demanded, for their renewal, fines for a total of 100,000 pounds. His tenants, mostly Presbyterian Protestants, offered to pay the interest of the fines, along with the rents, but to no avail. Town speculators eagerly entered the contest, some of them Catholics, ready to offer any sum to regain a hold on the land. Other landowners, such as Clotworthy-Upton, behaved like Lord Donegall.

The upshot was that Belfast speculators paid the rents and the fines demanded by the landowners, got the leases, evicted the former tenants and sub-let the lands. Thousands of families thus lost their homes. They founded the Steelboys movement and set out to maim cattle and destroy the properties of the new tenants who had replaced them. This outbreak of agrarian disturbances spread far and wide in Cos. Antrim, Derry, Down and Tyrone.

Very soon, many of the evicted tenants found their way to the North American colonies. The historical presence of Ulster Protestants in North America is of the utmost importance: their migratory flow was by no means insignificant in the early eighteenth century, and already at that time the oppression of landowners was being regarded as the main cause, along with the compulsion to pay tithes to the Established Church; a burden equally loathsome to Catholics and Presbyterians.

Emigration gained tempo in the early Seventies of the eighteenth century. Between 1770 and 1773, 101 ships sailed from Ulster ports, carrying over 30,000 emigrants. About a half of the American armies which, a few years later, defeated the Crown forces were, according to evidence submitted to the British Parliament, made up of Irishmen. Fifteen generals of George Washington were of Irish stock, as well as all the early U.S. Navy commanders. Four of the signatories of the Declaration of Independence were Ulstermen, while nine more had Ulster ancestry. The printer of the same Declaration, John Dunlap, hailed from Strabane; in 1784 he started the first daily of the USA in Philadelphia: "The Pennsylvania Packet."

A great many documents stress the link between the extortionate behaviour of Ulster landowners and the loss of the American colonies. Particularly significant among these are those of John Wesley, in an entry in his diary dated 15th June 1773; and a dispatch of 1802 to the "Irish Society" set up by the London municipality. Wesley wrote:

> When I came to Belfast I learned the real cause of the insurrection in this neighbourhood. Lord Donegall, the proprietor of almost the whole country, came hither to give his tenants new leases. But when they came they found two merchants of the town had taken their farms over their heads; so that multitudes of them, with their wives and children, were turned out to the wide world.

The Irish Society dispatch says, referring to Richard Jackson, a middleman of the London Clockworkers' estate in the vicinity of Coleraine:

> It is commonly reported in the country that, having been obliged to raise the rents of his tenants very considerably, in consequence of the large fine he paid, it [sic] produced an almost total emigration to America, and that they formed the principal part of that undisciplined body which brought about the surrender of the British Army at Saratoga.

Further corroborating evidence, concerning specifically Lord Donegall — whose estate in Co. Antrim alone included four baronies — can be found in a letter written to the same Lord, then in London, by his superintendent George Portis, dated 4th May 1773 (PRONI T. 1893), from which the flight of his tenants is plainly confirmed, though the writer is obviously at pains to soothe his Lordship by pointing out that the same thing is taking place in other estates as well.

While the tenants unleashed disturbances, went overseas in anger, waged war on their mother country and defeated her thus breaking the British Empire, the Chichesters, as firmly entrenched in Ulster as ever, went on, generation after generation, hatching plots to grab the Lough Neagh and Bann fisheries. It would be tedious to follow in detail the further feats of the succeeding Lords Donegall in the fisheries contest. The following cursory remarks will suffice.

When the ninety-nine years lease to Lord Massereene and his heirs expired, the claims of the Chichesters were forgotten. Legislation passed by the Irish Parliament dealt with Lough Neagh as a public good (*res communis omnium*), as it had always been. However, the plans of the Lord Donegall of the day were skilfully laid. The City of London had set up on the Bann, at the Leap of Coleraine, four salmon fisheries, which inevitably decreased fishing upstream. Without objecting to the Londoners' grant, Lord Donegall maintained that their fisheries were an infringement of his fishing rights in a pond on the Lough Neagh coast, in Co. Armagh.

Two suits failed in 1781 and 1784. A third attempt was made in 1787, carefully avoiding any hint that might raise questions on the 1661 patent, the only (spurious) property title for Lough Neagh and the Bann in Lord Donegall's possession. The point to be decided was merely whether the fisheries of the Londoners hindered the upstream migration of fish. This time the verdict went against the Corporations (1789). They appealed to the Irish Commons. It was the age of "Grattan's Parliament," and the Commons, swayed by the Anglo-Irish "Protestant Nation," rejected the Londoners' appeal.

In 1801, after the Union, the City submitted a new appeal which was upheld by the Lord Chancellor Fitzgibbon. The patents of the Chichesters, however, were not examined. Seemingly defeated, Lord Donegall had scored instead two important points in his favour: his claims on Lough Neagh had been deemed valid, and it had also been implicitly accepted by the Londoners that he held rights on the non-tidal Bann. The Corporations, in fact, had merely defended their rights on the tidal Bann.

Encouraged by this, Lord Donegall began to lease the river, and many agreed to pay the rent. The hold of the Chichesters on the fisheries became tighter and tighter, and it was only in 1868 that the City was rediscovered to be, on paper, the owner of the whole Bann. A lawsuit ensued, but despairing of overcoming long entrenched interests, the Londoners agreed in 1872 to pay his Lordship 2250 pounds *una tantum* for fishing rights in the non-tidal Bann, plus an yearly rent of 80 pounds, while the lucrative eel fishing was left entirely in the hands of the Chichesters.

The defeat of the London municipality prompted the Chichesters to try to suppress the public usage of Lough Neagh and the Bann altogether: a first lawsuit to that effect from 1873 to 1878 ended up in the House of Lords which passed sentence against Lord Donegall. A further attempt, in 1908, instead, was successful. In the meantime, however, the estate of the Chichesters had mostly been overtaken by Lord Shaftesbury and his heirs.

A sequence of lawsuits ensued, acted by the fishermen. After the most recent settlement of 1963, they are still acting to regain their traditional right to free fishing. According to Northern Ireland's regional bureaucracy, which is still operating at Stormont under direct British administration, everything is normal, as the fishermen must pay for a license issued by the same administration according to the Fisheries Act of 1966. Opposing this, the Fishermen's Cooperative maintains that the present situation is still the result of the usurpations of the Chichesters, since, as pointed out earlier Lough Neagh had previously been *res communis omnium.*

The losses sustained by the Crown in Ireland as a consequence of this "run" for Irish landed properties can be gauged, according to Healy, from the fact that in last years of the Union, immediately before the War of Independence, the Irish lands of the Crown yielded only £19,000 a year, while in England, where since the end of the War of the Roses in 1485 there had been no more massive confiscations, the income of Crown lands was £488,000, i.e. more than twenty-five times more. Allowing for the differences between the two countries in size and agricultural productivity, an imbalance of one to five, or even of one to seven or ten might be justified, but certainly not such a massive one.

I cannot, however, entirely concur with Healy when he maintains that the cost of the prostration of the Irish was borne by the British taxpayers while the profits fattened private speculators. The latter part of the statement is beyond question, but the British taxpayers did not always bear the burden of the conquest of Ireland. Cromwell, for instance, after his inhuman genocide, (*inter alia*, he had all the inhabitants of Drogheda and Wexford massacred, men, women and children) dispossessed all landowners east of the Shannon who were either Catholic or supporters of the King, and used the lands thus snatched (three quarters of all Ireland), to pay his soldiers and adventurers. In that case the bill was forced in the most direct and brutal fashion upon the Irish. Historically, the conquest of Ireland and the establishment of a colonial landownership based on outright usurpation go hand in hand.

The conquest had been cruel indeed, and, no matter what later "revisionist" historians might have contrived in their spasmodic attempts to say something new, we may take the judgement of Winston Churchill (1962, vol. II, p. 232) as a fair assessment:

> Above all, the conscience of man must recoil from the monster of a faction-god projected from the mind of an ambitious, interested politician on whose lips the words "righteousness" and "mercy" were mockery. Not even the hard pleas of necessity or the safety of the State can be invoked. Cromwell in Ireland, disposing of overwhelming strength and using it with merciless wickedness, debased the standards of human conduct and sensibly darkened the journey of mankind.

Yet it cannot be denied that Cromwell gave a powerful and decisive support, at a crucial moment, to the cause of Parliament and democracy. Any fair assessment of this historical period must acknowledge that if the United Kingdom, Ireland itself, the USA, and many other countries as well, enjoy an open society and an elected government, they must also thank Cromwell for this. Such are the ambiguities of history.

The conquest of Ireland was complete after the "Glorious Revolution," by the rise to power of William of Orange and his Irish campaign against James II Stuart, which was decided by the siege of Derry and the battle of the Boyne (1690), and concluded by the Treaty of Limerick (1691), which granted the Catholics at least freedom of worship. The Irish Protestant establishment were unhappy about the provisions of the Treaty which they regarded as too generous. Pretty soon the Irish Parliament began passing an extremely oppressive body of legislation, globally known as the Penal Laws, to discriminate brutally against the Catholic subjects.

Excluded from the voting franchise, from trade and professions except the medical one, from education, including being prohibited from sending their children overseas to study, Catholics could not hold public offices, be members of juries, carry arms and keep horses of a value above five pounds. The decreasing number of Catholic landowners were subject to legal norms aimed specifically at their wholesale dispossession. No Catholic was allowed to acquire land in any way except as a lease of up to thirty-one years. No Catholic could obtain land from a Protestant through marriage or inheritance, nor could he dispose of his land by will; the whole estate was automatically divided among all his children, unless one of them "conformed" to the Anglican Church, in which case he inherited the whole estate. If a son conformed while his father was still living, the latter became a mere life tenant. It is estimated that between 1703 and 1788 about 5,000 Catholics of the aristocracy and the middle classes conformed in the whole of Ireland; a rather small number, especially for such a long period of time, and, since the refusal to conform in most cases was tantamount to the ruin of the family. By 1770, when the Penal Laws began to be applied with less severity, the Catholics were left with but 5% of the land in the whole island.

James Boswell, the biographer of the great Doctor Johnson (certainly not a radical but rather a well known supporter of "subordination" and the establishment), relates that the Doctor

had a great compassion for the miseries and distresses of the Irish nation, particularly the Papists; and severely reprobated the barbarous debilitating policy of the British government, which, he said, was the most detestable mode of persecution. To a gentleman, who hinted such policy might be necessary to support the authority of the English government, he replied by saying, "Let the authority of the English government perish, rather than be maintained by iniquity (.....)."

What the Irish suffered under the burden of the Penal Laws for over a century, and what they lost in terms of suppression of their intellectual abilities is beyond reckoning. The position of the Catholics was particularly depressed in Ulster due to the large presence of a numerous Protestant population which had controlled the land almost entirely from a very early date. For a large majority of the people any kind of respect for the law vanished: in the popular imagination the image of the outlaw became that of a hero and a friend of the people — very much as Robin Hood had become in Medieval England for the Anglo-Saxons under Norman rule.

The fact that the Penal Laws were enforced sometimes with great inefficiency did not take away the smart of discrimination and merely lessened the respect for the law even further. The sheriffs of the Counties were local landowners: unpaid amateurs assisted by utterly embrionic police forces. Even when the authorities succeeded in apprehending an important member of the Catholic clergy such as John Fottrell, provincial father of the Dominican Order, nothing happened, though at the time of his arrest in Thome, Co. Derry in 1739, he carried a written report on the conditions of the many Irish Dominican friaries. The document was seized, but Father Fottrell fairly soon escaped and took refuge on the continent. Only the land-grabbing part of the Penal Laws was enforced with true ferocity and devastating effects.

During the eighteenth century the Irish Protestants, who were firmly in power, exploited their territorial conquest causing the final disappearence of the traditional Gaelic society. The last remnants of tribal Ireland vanished when the Act of Resumption of 1700 returned to the State all lands confiscated or granted after 1689, and ordered them to be auctioned off. Thus, the last traces of the old communalistic property of the clans disappeared along with the economic mainstay of traditional society. However, the suppression of clan particularism contributed to the national unification under the influence of the Catholic Church, the only beacon left to the Irish. This, in the long run, could but undermine the foreign domination.

In another sense the conquest was undermined from the outset. It had invariably been a penetration, and therefore a development, of the "hollow frontier" type. The waves of the various conquerors, from the Vikings onwards, had penetrated deeply, playing havoc and imposing their domination more and more completely, but they had become dependent upon the local manpower. It was therefore not a very dynamic kind of conquest and development, based upon the control of strategic points and the build-up of huge landed properties, not unlike the cases of South Africa, Mexico and Peru. Only in Eastern Ulster and in some other areas of the same province was the conquerors' presence so strong as to resemble, to some extent, that of the whites in North America.

The social and spatial imbalance between the Protestant élite and the majority of the population reached abysmal levels. In most of Ireland, the majority of Catholics became either cottiers or tenants-at-will, under landowners almost exclusively Protestant, who imposed rents so high as to leave no margin above bare survival levels. In 1770, the Lord Lieutenant wrote to King George III:

> I hoped to be excused for representing to His Majesty the miserable situation of the lower ranks of his subjects in this Kingdom. What for the rapaciousness of their unfeeling landlords and the restrictions on their trade, they are amongst the most wretched people on earth.

All eighteenth century observers agree in depicting a wholesale social degradation, though with the exception of many areas of Ulster, where the tenants were usually also Protestant (mostly Presbyterian) and enjoyed a treatment that, though still extortionate, was less inhuman. The "Ulster custom," whereby tenants were more difficult to evict and enjoyed a right to compensation for improvements made on the estate, was universally envied by the subject rural population in the rest of Ireland, though, as we have seen, even that was no defence against particularly greedy landowners.

Protestants did not refrain from making their dominion felt in the harshest way. In 1792, an inhabitant of Sligo, Co. Donegal, gave Wolfe Tone "a most melancholic account" of the depression and insult under which Catholics of that town were labouring:

> Every Protestant rascal breaks their heads and windows for his amusement (.....) the Catholic spirit is quite broken.

Many who would have never sought dominion if it had not been prepared for them by others, wished to preserve their privilege and even the last of the Protestants, if he had nothing else to boast about, found it satisfactory to think that, at least, he belonged to the "master race." Many white South Africans in the heyday of their power doubtless felt the same way. In contriving the Penal Laws, the Irish Parliament acted more ruthlessly than the metropolitan British power; likewise the South African Boers were more hostile to the blacks than the far-away British government.

As in South Africa, towards the end of the seventeenth century Huguenot refugees settled in Ireland. They caused no problems; their presence proved to be a considerable economic asset, on account of their advanced techniques, as we shall see in dealing with the development of Belfast (Chapter 7).

On the other hand, the massive immigration of Scottish Presbyterians, especially into Ulster, at the time of the "Glorious Revolution" and shortly afterwards, was felt by the ruling Anglican élite as a new threat. Anglicans in Ireland, already a puny minority in comparison with the Catholic masses (though politically powerless), seemed in a state of siege. In an attempt to entrench their power, the Anglican élite caused the Test Act to be passed in 1704: a law against Catholics and Presbyterians alike. As a consequence, in the same year the Mayor of Belfast, David Buttle who was a Dissenter, having refused to conform was obliged to relinquish his post.

Until the end of the eighteenth century a Protestant man who married a Catholic woman who did not conform within one year of the match, sank into the condition of "papist," was stripped of all rights and became, in the eyes of the other Protestants, more odious than the true Catholics. Mixed marriages, however, continued to take place from time to time, and this kept the dominant group from become entirely closed.

There was also resentment of the Protestant Irish against Britain, occasioned by the economic discrimination aimed at Ireland as a whole. This policy was particularly damaging to the economic élite (landowners, manufacturers, merchants) and thus above all to the Protestants, but also to the well-to-do Catholics. Many trade restrictions had been introduced by the English Parliament during the seventeenth century. Every new wave of settlers who made Ireland their home became the target of colonial treatment by the English metropolitan power. Ireland's dependency was thus wholesale, and all Irishmen who were not threatened by hunger had, in the eighteenth century, irrespective of their religion, some common ground for solidarity — at least potentially — against the colonial power.

Ireland attracted unscrupulous speculations from individual British subjects. Between 1722 and 1724, the scandal of Wood's halfpence broke out: an attempt by an enterprising Englishman to coin, under license from the British government, a huge number of worthless brass coins destined exclusively to circulate in Ireland. The economic consequences of such a flood of bad money would have been disastrous. The venture foundered due to the storm of protest which arose in the island.

Jonathan Swift, the Dean of St. Patrick in Dublin, published on that occasion, his *Drapier's letters* (1724), a volume of strong satires. Later (1729), taking inspiration from the dreadful famine of 1726-29, due to several consecutive bad oats harvests, which caused the death of about 400,000 people in the whole of Ireland Swift attacked in an even more savage satyrical vein the society of the time in his famous pamphlet: *A modest proposal for preventing the children of the poor people from being a burden to their parents or the country and for making them beneficial to the public.*

Irish Protestant intellectuals had good reasons to complain. The highest government posts in Ireland were given not to Anglo-Irish but to Englishmen only. Throughout the eighteenth century, no Irishman was given the Anglican Archbishopric of Armagh, to which the Primacy of Ireland was attached. Funds of the Irish Exchequer were regularly used to pension seasoned English statesmen, even when the Irish government was deep in debt. Nothing was being done to stem the corruption of the Irish Parliament, bitterly but unavailingly denounced in his *Political Addresses* (1748-49) by Charles Lucas, who was one of the many disaffected Irish intellectuals.

While disaffection was growing among Protestant intellectuals, the fear of the defeated Catholics subsided, and — after the battle of Culloden in 1746 — that of a return of the Stuarts utterly vanished. The condition of Catholics, or rather of their small well-to-do class, improved somewhat when the Pope, in 1776, decided not to uphold the claim of Bonnie Prince Charlie to England's throne, as any hope of a new Stuart Restauration was evidently lost.

Moreover, the British ruling circles began to worry about the drain of military manpower brought about by the exclusion of Catholics from service, just at the time of frequent wars such as in the eighteenth century. Not only did they not serve the King, but they went to the continent to serve his potential enemies, building up mercenary troops known in the whole of Europe as "the wild geese."

So, while the attitude of the British government was slowly changing, the Irish Protestants, then in the heyday of their fortune and power, began, in the second half of the eighteenth century, to regard themselves as "the Irish Nation." Some kind of solidarity developed between the Catholics and Protestants due to the common economic interests *vis-à-vis* the privileges of

British merchants, and to the political and social discrimination under which both Catholics and Dissenters were labouring.

Leading Protestant personalities such as Henry Grattan and Wolfe Tone stood for Catholic emancipation and a common Irish nationality, while large numbers of Catholics became supporters of the existing social order, and joined the voluntary militia recruited for the suppression of agrarian crime committed by their fellow Catholics. Organised in the Whiteboys movement, they maimed cattle and intimidated landowners to dissuade them from enclosing their lands.

Towards the end of the eighteenth century, the Penal Laws, already applied with considerable slackness, were gradually dismantled. A new form of oath reserved for Catholics, which did not compel them to abjure Catholicism was introduced in 1774, the purchase of land by them was permitted (1780), the Test Act was abolished, and two Catholic Relief Acts were passed in 1792 and 1793, of which the second gave them voting rights, but not yet the right to be elected. These concessions were granted by the British government who were concerned about the spread of revolutions throughout the Western world, and obliged to engage the garrison of Ireland overseas. The Irish Parliament, instead, was averse to these innovations: this confirms once more the harsher hostility of the peripheral ruling élite towards its own subjects in comparison with the metropolitan élite whose horizons were, of course, not so narrow.

A new patriotic exaltation galvanized the Irish, or at least that minority which did not have to worry too much about their daily bread, or rather about their daily potatoes. The movement of the Volunteers was formed. The repeal by the Irish Parliament of the laws keeping it under the sway of the British one was, however, the most revolutionary measure: under the idealistic guidance of the "patriot party" headed by Henry Grattan, Henry Flood and the Earl of Charlemont, in 1782 the Irish Parliament repealed the Poynings' Law of 1495 and the 6th of George I of 1720. Between 1778 and 1780 free trade was achieved. The same Parliament, on the other hand, refused to introduce the democratic reforms suggested in 1783 by the Volunteers, thus upholding Protestant privilege and anti-Catholic discrimination.

This Parliament was little representative even of Protestant opinion. The Irish Commons had 300 seats, 64 for the Counties (2 for each County) and 236 for the 109 boroughs, many of which were in fact rotten boroughs. Sometimes only the town councils, made up of 12 Aldermen for life, elected the representatives of the borough. Irish M.P.s themselves were nearly unmoveable before the Octennial Act of 1768, which had made compulsory that elections be held at least every eight years.

Before this law, elections were held only when the King died. Nearly all borough seats were, in one way or another, controlled by landowners: according to an inquiry made by the Volunteer movement and submitted to the same Commons in 1783, half the seats of that Parliament were controlled by only 57 great landowners. Cases such as that of St. Johnstown, Co. Donegal, which, having 150 inhabitants, chose two representatives, the same number as Belfast which had 15,000 inhabitants, were fairly commonplace.

The British government watched the outbursts of Irish ebullience in the Dublin Parliament rather passively: the strategic vulnerability of Ireland and its ideal situation, had it fallen prey to an enemy, as a base for an attack upon Britain itself, made counsels of caution prevail in London. Moreover the large population, which had surged from approximately half a million after the partial Cromwellian genocide to about four million in 1775, made Ireland an important source of recruits.

With the outbreak of the French Revolution, not a mere national one as the American, but a social one, posed a more severe threat to the establishment. France was also geographically closer to the British Isles, and that was an additional factor to worry about. Wolfe Tone published in 1791 his *Arguments in favour of the Catholics of Ireland,* and in the same year he founded the United Irishmen.

Some people were even more radical, at least in theory, than Wolfe Tone himself. One of these was William Drennan of Dublin who, in a letter dated 3rd July 1791, suggested to his brother-in-law Sam McTier of Belfast (PRONI D.591/303), that the parade of the Volunteers of Belfast, already planned to celebrate the second anniversary of the storming of the Bastille, might have been a good occasion for a show of friendship towards the Catholics. The suggestion, however, was opposed by many Belfast reformers. Wolfe Tone himself disagreed with Drennan, and when the United Irishmen were founded three months later in Belfast, their official ideology was rather hazy on Catholic civil rights.

The United Irishmen, faced with increasing government repression, became a clandestine movement in 1794. The following year Wolfe Tone left Ireland and went first to the USA and later to France, where he sought to persuade the revolutionary government to mount an armed invasion to free Ireland. A first aborted attempt by the French took place in 1796.

In the meantime, violence against the Ulster Catholics was rising. Low class Protestants had set up a terrorist organisation known as the Peep O'Day Boys. The Catholics had formed the society of the Defenders, particularly strong in Southern Ulster, a group often regarded as primitive, sectarian, reactionary, essentially agrarian and local, but which appears, under more careful scrutiny, a radical, sophisticated movement, politically up-to-date, unusual in having a leadership and an ideology independent from the middle classes, effective just for being closely in touch with the local realities of Ulster Catholic life (Wheelan). The Defenders clashed several times with Protestant gangs. After one more clash, which ended in a Protestant victory, the Orange Order was set up and the hunt for Catholics continued. Tens of people were murdered, thousands of refugees were forced to flee to Connacht, especially to Co. Mayo. The slogan of the Cromwellian deportation, "Hell or Connaught," was revived at that time.

Whiskey played an important part not only in the unrest, but also in political trials, as in the case of William Orr, from Farranshane, Co. Antrim. The man was hanged in October 1797 for administering the oath of the United Irishmen, which had been outlawed, to two soldiers. Rumour had it that in order to ease the verdict, some members of the jury had been made drunk with whiskey. Moreover, one of the two soldiers who were witnesses against the accused, was insane. Nevertheless the government decided to carry out the execution. Orr gave a speech from the scaffold proclaiming his innocence and printed copies of a statement of his were distributed to the crowd (PRONI T.2627/4/92), while the executioner was proceeding with his grisly task. This death, along with the brutality whereby, in the same year, the troops of General Lake seized the weapons held by civilians, inflamed the Protestant population of Ulster.

Ulster's Catholics, instead, were too concerned with the violence and intimidations under which they were labouring, to contribute heartily to the agitation, while in the other Provinces they played a far more active role. Thousands of Catholics in the northern Province deserted the uprising as soon as the clash came to a crunch, whilst their fellow Catholics in the Royal forces — several regiments were almost entirely manned by them — fought to the end. The disarming of Ulster had deprived the insurgents of their only supply of weapons worthy of the name. In the spring of 1798, the United Irishmen were easily defeated. The government reacted with ruthless efficiency, well supported by informers from whom information was often extracted by means of torture.

Outside Ulster, the insurrection scored a local success in southern Leinster, only to be crushed a few days later at Vinegar Hill. In this area, which was for a few days out of government control, the insurgents went roughshod settling accounts with local landowners. This episode estranged the Catholic and Protestant Irish even further.

The lack of coordination with the French landings also fatally affected the outcome of the struggle. Napoleon's troups sought to invade Ireland two more times in 1798, but failed, due to adverse weather and a highly effective British reaction. In the last of these attempts, a French expedition sailed into Lough Swilly, in Donegal, but was defeated by the British fleet before it could land. Wolfe Tone, who took part in the expedition, was apprehended and sentenced to death for high treason, but committed suicide in jail before the execution.

The demise of Ulster as a spearhead of republicanism was rapid and final. The northern Province gave no support to the Irish struggles in the nineteenth century. Though interconfessional republicanism was over by the end of the eighteenth century, it had paved the way for the future Catholic republicanism.

The British government, deeply worried about the Irish unrest — a thorn on its side just at a time when the French threat appeared extremely serious — reacted not only with military and police repression, but also with political measures: the abolition of the Irish Parliament and the Union of the two Kingdoms. In 1800, effective from the 1st of January 1801, Ireland became a part of the United Kingdom, under a common Parliament in which only Protestants could hold seats.

In spite of its peripheral location, Ireland felt the shock waves of the European revolutions and, in the nineteenth century, became a focus of revolution in its own right. The golden age of the Protestant Ascendancy was on the wane, while the scarce flexibility of the British government, the rigid socio-economic structure, the dependence of the lowest rural classes on the potato and the swift demographic growth, were all paving the way to that tragic event, the Great Famine; a point of no return for the domination of the British metropolitan Centre with its sub-imperial appendix made up of the privileged Protestant (especially Anglican) ruling élite.

The dependent status of Ireland was apparent from every viewpoint — political, economic, cultural. The largest towns were but "gateways" of imperial commerce and administration: in a scarcely industrialized country, urban life had but little else to live upon. Towns grew more slowly than the Irish population as a whole; only Dublin and Belfast were exceptional in this respect. Between 1821 and 1841, the former grew by 25.2%, while the latter soared by 133.3%. In some towns, more than one third of the population was made up by workhouse inmates. Evidence of the persistent demographic weakness of towns is shown by the expectation of life there, estimated by the 1841 census (about 28 years for males and 29 for females), which was considerably lower than in the countryside (males 34, females 33).

There were nevertheless some hints of economic development. The first Irish railway line, the Dublin-Dun Laoghaire, was opened in 1834. It was followed in 1839 by the Belfast-Lurgan which was extended to Portadown in 1842. The longest railway line built before the Great Famine was the Dublin-Drogheda (51 km) which was completed in 1844. Roads and canals were hard hit by the competition of the railways, especially the canals, whose flexibility was extremely low. Canals had been introduced in Ireland in 1741: the first one had linked Newry to the sea. The Ulster Canal which connected the Erne and the Blackwater was completed in 1842 when the threat of railway competition was already looming large.

Eastern Ireland, taking as a divide an ideal line from Derry to Cork, was far more developed, while the western part of the island lagged behind. The critical factor of this imbalance was, and still is, the greater accessibility of the East from Great Britain. This had already brought about an intensive anglicization and a stronger Protestant presence, while in the West the Gaelic language was still strong and the dominance of Catholicism nearly unchallenged. The most favoured Province was Ulster, especially its eastern part, which was becoming polarized around the buoyant town of Belfast.

Functionally, neither Dublin nor Belfast belonged to Ireland, but rather to the dominant island. If Dublin was a projection of London, Belfast was a projection of industrializing Britain. Dublin could not compete with London but rather, as a dependent subcentre, and a main gateway and major administrative centre (no longer a politically autonomous centre), was a vital link in the imperial chain of command and cooperated in strenghtening London's hold on Ireland. Belfast, where the main textile activity was based upon linen, in no way competed with the textile towns in Britain, since these worked other fibres: cotton (Manchester, Nottingham, Leicester, Glasgow), wool (Leeds), or silk (Derby). Linen factories were rare in Britain, mostly to be found in Manchester, and the British linen interest was not strong enough to overcome the Irish linen industry.

It is very likely that the comparatively less disadvantaged position of Protestant tenants in Ulster might have favoured the necessary capital accumulation for the take-off of industrialization in the northern Province. Some revisionist historians (e.g. Cullen) have challenged this interpretation, stressing that cottage weaving was also to be found in Catholic areas. This, however, fails to explain why the linen industry became concentrated in Protestant hands, whereas the small independent weavers — the Catholic ones, and obviously a great many of the Protestant too — vanished from the scene.

Cullen maintains that the restrictions to trade which hampered the industrial initiatives of the Irish, except the linen industry and the brewery, were not aimed at hurting Ireland: a very surprising statement, to say the least. No scholar worthy of the name ever suggested that the commercial restrictions were a part of an evil conspiracy against the Irish (or against the American Colonies). Protectionism was used when competition from Ireland or elsewhere was perceived as a threat by the British commercial interest, while the market was liberalized when the British position was so strong as to be able to overcome competition without political support. The damage to the Irish economy was just an added consequence, no less serious for being unintended.

Not surprisingly, the economic situation did not favour law and order. Between 1800 and 1845, nearly every year the London Parliament enacted Coercion Bills: these were aimed at checking the movements of individuals and the right to carry arms, and allowed heavy jail and transportation sentences to be passed on people regarded as dangerous. Australia received a steady influx of Irish convicts, often guilty merely of minor offences or entirely innocent. The accused had but few opportunities to defend themselves, because they did not understand the procedure, and therefore had no idea of their rights, and also because they often had just a smattering of English or none at all.

The uprising headed by Robert Emmet in Dublin in 1803 was easily crushed. However, the fears it excited, along with the ever present agrarian crime, kept the government in never-ending alarm.

Adaptive measures were also taken, aimed at satisfying the rising Irish middle classes. Such was the foundation of the Kildare Place Society (1811) which, with the financial support of the government, offered an interconfessional education. The innovation, however, met with scanty success, due to the opposition of the Catholic Church, by then an acknowledged power. Tangible evidence of this acknowledgement was the setting up of a government subsidized Seminary, the Maynooth College (1795), since access to Universities and Seminaries in France and other European countries had become difficult during the French Revolution.

In the meantime, however, the Catholic refugees from Ulster, under the threat of Protestant violence which followed the foundation of the Orange Order in that same year, had become agents of propaganda against the government which were far more effective than the Protestant intellectuals had ever been.

Also the way in which the British government (which had moreover to take account of the even greater reluctance of the Irish Protestants) had eventually come to make concessions was so fragmentary and reluctant as to pave the way for future conflicts. After the Catholic Relief Act of 1793, Catholics could vote, but did not yet have the right to sit in Parliament. This meant that the Catholic aristocrats met with serious difficulties in their attempt to regain their traditional leading role. Moreover, they were so excessively compliant as to be ready to accept, in 1808, a veto right by the government on the appointment of Catholic bishops. In time, the rising Catholic middle classes (lawyers, merchants) supplanted them. They found a leader in Daniel O'Connell, whose strenuous efforts succeded in achieving Catholic Emancipation in 1829.

In fact, the premises for emancipation had been in existence some time. In the darkest Penal Age, Mass was usually said in the fields, against big outcropping stones (Mass rocks, in Gaelic *scáthláns*). Later Mass houses grew greatly in number. The young Catholics, who according to the Penal Laws should have been doomed to illiteracy, were educated in the open countryside, not only in the three R's but also in Latin. Friaries survived in out of the way streets in the towns, under the mask of private houses.

As shown by Wheelan, the Catholic Church had already begun reorganizing in the eighteenth century from a strong Catholic core in central Munster and southern Leinster. The diffusion had been slow only northwards, into northern Connacht and Ulster. In the latter Province the historical experience of the Catholics had been particularly degrading: the highest positions they could achieve were those of shopkeeper, hostler, butcher or schoolmaster. The "Mass houses" were seldom to be seen there, so the Eucharist was celebrated, more often than elsewhere, at the *scáthláns* by itinerant priests or friars.

Catholicism made up a powerful indigenous culture in the core, while in the North and the West a more popular form prevailed, in which archaic beliefs and superstitions, though condemned by the Church, still survived. Geographers, says Wheelan, have often commented upon the existence in Ireland of an East-West dichotomy, while more precisely there exists a dualism between South-East and North-West. The Catholic core from central Munster to southern Leinster, coincident with the traditional Counties of *hurling*, was to become the heartland of the future Republic. The other important core — Ulster east of the Bann — made up the heartland of future Northern Ireland.

After 1793, the Irish Catholics managed to exercise a considerable, sometimes decisive, influence on elections in many county seats, while very seldom they could control the borough elections. With the Union, the Irish seats of the new common Parliament of the United Kingdom of Great and Ireland had been reduced from 300 to 100 by the abolition of 200 borough seats. This was tantamount to giving the Catholics greater power, albeit still indirectly.

Together with the Catholic Emancipation Act, two "wings" were passed: i.e. two laws specifically aimed at dampening its effects by lessening the strength of the Catholic electorate. One of them dissolved O'Connell's Catholic Association (which had already disbanded), the other raised the franchise from 2 pounds (the well known 40 shillings franchise) to 10 pounds.

CHAPTER 3

FAMINE AND AFTERMATH

The Protestant élite hoped that emancipation, such as that which had been doled out to Catholics, would weaken the political strength of Catholicism and create stronger ties between Great Britain and Ireland. The Catholic élite, headed by O'Connell, hoped instead that emancipation would mark the outset of social redemption for Catholics. Both were wrong, and the worst was still to come. During his travels in Ireland, around 1835, Alexis de Tocqueville heard from a peasant this bitter remark: "Emancipation has done nothing for us. Mr O'Connell and the rich Catholics go to Parliament. We keep starving just the same."

Neither O'Connell nor his close abettors, all of them basically conservative in outlook, and not even the radicals of newly hatched "Young Ireland," belonging to the urban middle classes such as John Mitchel, son of a Presbyterian minister from Ulster, gave serious thought to the problem of the dreadful imbalances and social tensions. Poverty in town and country attracted comments from all foreign travellers: their impressions were perfectly in keeping with those from eighteenth century observers.

Supporters of the fashionable "revisionist" trend are at pains to stress that the relationships between landowners and peasants were not always marked by exploitation and conflict, and point to the need to take account of not only contemporary observers, but also of estate papers. This is quite an acceptable proposition. Let us therefore look at the papers of the third Marquis of Downshire (1809-45), edited by Maguire (1974). There we find a letter addressed to the nobleman, dated 31st August 1815, from one of his agents, named James Brownrigg, from Edenderry (PRONI D. 671/C/237), in which we read:

Here is no Yeomanry, — no agricultural capitalist, no degree between Landlord & Labourer, the words "peasantry" and "poor" synonimously employed: this is a brief but true picture of Ireland. The Landlords have generally acted with the most extraordinary want of foresight in the Leting (sic) of their Lands — by subdividing their farms into small divisions, and leting (sic) them to the very lowest class of the people, at high Rents.

These peasants exhausted the ground by sowing wheat, under the protection of the Corn Laws, even when the land was unsuitable and had been previously exploited only for pasture. Although well informed on the situation, it did not prevent the Downshires, in the person of the fourth Marquis, from behaving in a remarkably hard manner when the great Famine broke out.

The causes of the widespread poverty were manifold — demographic, economic, social, and political — but all, in one way or another, were linked to social and spatial imbalances fuelled by the colonial situation.

A factor of the highest importance was the swift population growth, which seems to have undergone a spurt after about 1780. The reason for this was not so much a decreasing deathrate, but rather a rising birthrate. It is well known that a high birthrate can occur in times of economic development and income growth when people are encouraged to marry earlier, as occurred in England at about the same time during the first Industrial Revolution.

But in Ireland the position was utterly different. The wretched poverty of so many rural families made the young irresponsible. Any change was regarded as a relief. The promiscuity in the overcrowded huts prompted the Catholic Church to encourage early marriage, for moral reasons. "Labourers get married under the impression that they cannot make their condition worse than it is," and "At any rate, if the worst comes to the worst, the wife can take to begging," were two widespread comments of the time. It was doubtless a population increase linked to obvious social pathologies.

The rigidly entrenched privilege of the Protestants went hand in hand with a lower birthrate. Like most privileged classes, this group was bent on an individualistic enjoyment of its privileges, and was therefore wary of begetting children thoughtlessly. Exactly the opposite behaviour of the Catholic majority led to a surge causing a strong population imbalance. The inaccuracy of estimates and censuses of the past makes it advisable to leave out details and concentrate upon long term comparisons which show both a trend of lively population growth and a steady increase in the percentages of Catholics in all Provinces (see Fig. 3.1).

The 1841 census shows that Ireland had nearly 8.2 million inhabitants, the highest number ever officially recorded, and it is safe to assume that the population kept growing until 1845, the first year of the Great Famine,

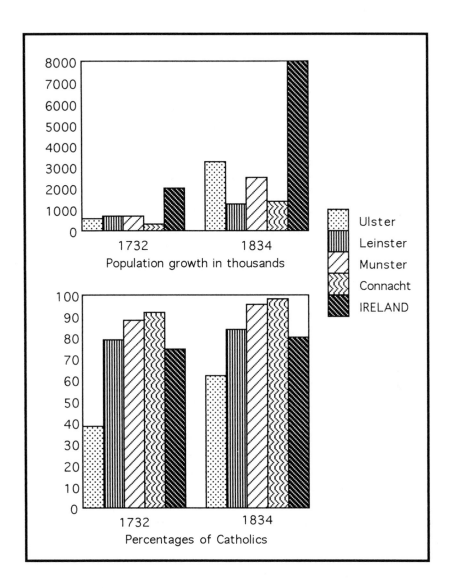

Fig. 3.1 — Population growth and percentages of Catholics by Province. (Sources: for 1732 *An Abstract of the Numbers of Protestant and Popish Families in the several Counties and Provinces of Ireland taken from the returns made by the Hearthmoney Office in Dublin, 1723-33*, Dublin, 1734; for 1834 *First Report of the Commission on Public Instruction, Ireland*, London, 1835, completed by data from Beaumont, 1839, and McCulloch, 1839).

reaching perhaps 8.5 million. According to recent estimates (Lee, in Goldstrom & Clarkson), the census figures of 1841 were underestimated by nearly 3%, and in that year the population was probably above 8.4 million; in 1845 it might have reached 8.7-8.8 million. If these revised figures are correct, then, the mortality caused by the Famine might have been worse than commonly thought: not one million but nearly one million and a half.

The fast population growth was partly due to the vested interest of landowners to divide their properties into small plots, due to the fact that between 1793 (Catholic Relief Act) and 1829 (Catholic Emancipation Act) the forty shilling tenants enjoyed the franchise, so that it was profitable for a landowner to have many of them in order to control as many votes as possible. Such a control was very easy because the vote was to remain public until the passing of the Ballot Act in 1872. The largest landowners employed men, called drivers (the same term used for men who drove slaves or cattle), to lead the tenants to the polling place and make sure they voted as directed.

In 1829, when the forty shilling tenants were disfranchised, the political interest of landowners to have many tenants ceased, and the economic interest prevailed: this meant a drive towards bigger plots to be used for commercially oriented pastoral farming. Tenant evictions increased dramatically. Evictions had already become frequent due to the economic crisis which followed the end of the French Wars. Precisely in that period, in 1816 and again in 1820, laws to make ejections easier were passed: they allowed the seizure of harvests including that of potatoes, which were usually the only means of survival for the peasants and their families.

On the other hand, the Penal Laws obliged Catholics to inherit through partible inheritance only. Although the penal age was over, its aftermath was by no means finished and the consequences continued to be felt, especially in the West which was still largely Gaelic and already lagging economically due to structural weakness and poor accessibility. In fact, when the Famine broke out, eastern Ireland, better connected to Great Britain by a network of interests and relations, suffered less disastrously. Eastern Ulster suffered the least, especially its Protestant population, who had access to alternative employment in the linen industry.

Landowners were no help to the starving. The legal implications of landownership were often felt as something unconnected to the moral duties of a landlord towards his tenants, whereas in England such duties were far more keenly felt. Ireland still resented the effects of invasions which had never been assimilated. In most of the country, with the exception of parts of Ulster (especially Cos. Antrim and Down) where the tenants too were

Protestant, and with the further exception of a few properties still held by Catholics, a Protestant landowner lorded upon "papist" tenants and cottiers. He seldom interacted with them except through agents, who often were Protestant too.

Land was usually seen by the owners as a mere investment. Holdings were sometimes mortgaged but this was not always sufficient to yield an income that might support the standard of living — often too expensive and pretentious — of the family. Maria Edgeworth, in her novel *Castle Rackrent*, published in 1800, admirably depicts the profligate life of the great landowners up to two generations before the Great Famine.

The social structure in the countryside was strongly stratified: great landowners were at the top; below them there were many smaller ones (some not particularly rich), then, on a lower level, the superintendents (also called agents); then the tenants (many of whom were very poor, while others could be reasonably well-off, especially in Ulster): some of these acted as middlemen, subletting land; the lowest order was made up of the cottiers. Moreover there were, especially in estate towns or other rural towns, artisans, hostlers, divines and other rural inhabitants not engaged in agricultural pursuits.

How and when the potato came to Ireland and became an important crop, is not precisely known. It is possible that because it grew under ground and was not yet generally known in the rest of Europe, the new crop became more and more common because it had a higher probability of escaping the plunder and the deliberate scorched earth policy practised by the invading armies. Later, it became a staple food for the rural poor in the depressed social conditions caused by the conquest.

The high productivity per acre of the potato was a potent factor for population growth. Cottiers, and also many small tenants and subtenants, seldom had occasion to handle money as even rents and salaries were reckoned in "potato-units," according to the so-called "potato truck system." For these people to be deprived of potatoes meant to be without food and without money at the same time.

In 1844, there were 98 workhouses in Ireland, which harboured 86,000 people. These drab and forbidding establishments were going to be pretty soon under siege by starving crowds. To Irishmen, even before the Famine, this system was a source of grievous suffering: many who sought shelter there, wished only for some temporary relief to get over a difficult period but, in order to be admitted, they had to prove themselves utterly destitute and were thus compelled to forego any possession of theirs for good. Moreover, the family was broken up: man and wife, parents and children, all were housed separately. This system continued to be in operation throughout the remainder of the nineteenth century.

The *Third Report of the Irish Poor Commissioners* (1837) is of the greatest interest. The behaviour of the landowners, explains the report, had raised among the peasants an "uncontrollable spirit of rebellion;" "landowner and peasant lived in a perpetual state of reciprocal fear." These are official British parliamentary sources, which, in spite of "revisionist" opinion, can be neither neglected nor easily discounted. Is it possible that they give an exaggerated picture to favour a given political faction? But then what of the foreign travellers who had to please no political master, and who paint exactly the same picture of destitution and conflict?

It is obvious that the Great Famine hit a society already sunk into a deep crisis, whose conditions are even more explicitly illustrated by the *Inquiry into the state of the law and practice in respect to the occupation of land in Ireland*, published in 1845, the very same year in which disaster broke out (Fig. 3.2). Insufficient salaries obliged cottiers to rent puny plots where they grew potatoes of the grossest variety. In many areas the potato was their only food, water their only drink, their huts seldom provided protection against the weather, a bed or a blanket were rare luxuries, a pig and a heap of manure their only property. When a cottier was sued because his pig had trespassed on a field, or because his manure heap was obstructing a road, it was often found at the trial that he had neither the means nor space to build a sty or to prevent his manure from invading the road.

The law courts, on the other hand, were manned by the landowners. It was a typical *ancien régime* situation, worsened by the consequences of the conquest. But for the threats of the secret societies, the landowners were unchallenged masters. The tax for poor relief was practically paid by the poor themselves: if a tenant applied for relief, the landowner used to reply that his rent had to be raised by the same amount to get money for the relief. Many tenants, and all cottiers, could not afford to buy glass, even of the grossest quality, due to the high customs duties on glass articles: a circumstance — says the report — fraught with serious health hazards.

The suffering of evicted tenants and cottiers, as it appears from the report, was dreadful: sunk into abject poverty, they were thrown into "any kind of vice and iniquity to survive," and in spite of that, a great many of them actually starved. Evidently, even before the Great Famine, there was already a widespread, endemic famine, for which the British Commissioners do not hesitate to lay the blame on the social conditions. Famine had already visited Ireland a great many times for centuries, nearly each generation, so as

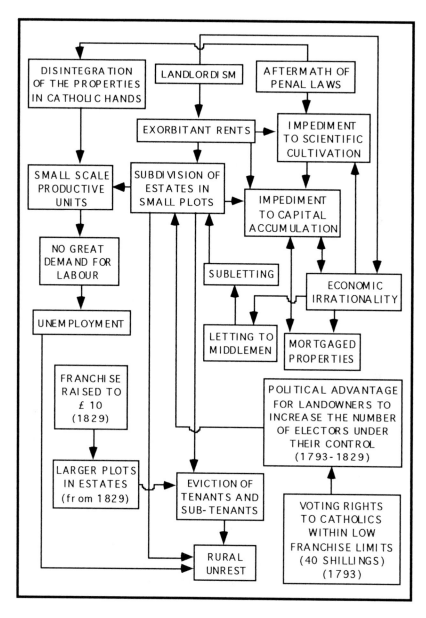

Fig. 3.2 - Social conditions in the Irish countryside from the *Third Report of the Irish Poor Inquiry Commissioners*, London, H.M. Stationery Office, 1837, and the *Report from Her Majesty's Commissioners of Inquiry into the state of the law and practice in respect to the occupation of land in Ireland*, 3 vols., Dublin, H.M. Stationery Office, 1845.

to give rise to the grisly legend of the Man of the Great Hunger, the *Fear-Gorta*, a dreadful skeletonized spectre who knocked at doors, begging for food.

The immediate cause of the potato blight was a microscopic fungus, the *Phytophtora infestans*, which thrives in extremely moist ground. Environmental conditions were made favourable by the heavy rainfall of 1845 and the ensuing years. The Famine lasted until 1849, or, according to some authors (Kinealy), even to 1852. The tubers became rotten and yielded an inedible blackish poultice.

Again, these biological processes were a part of a broader three-sided interaction with Britain and the New World in which Ireland was the weakest partner. Human migrations across the Atlantic had brought forth the transfer of other organisms: the potato had thus come to Ireland from the Andes region, and the parasyte fungus *Phytophtora* followed it from North America. A few decades later another pest, in the form of an insect, the potato beetle (*Leptinotarsa decemlineata*), also arrived from North America, though with somewhat less dire effects than the potato blight.

The first victims of the disaster were the cottiers' pigs, slaughtered because there were no longer potatoes to feed them. Then human casualties began to occur. In addition to the Famine there was the explosion, in epidemic form and often with deadly effects, of a great many illnesses: scurvy due to scarcity of vitamin C, dysenteria, cholera, typhus, ophtalmy leading to blindness due to the scarcity of vitamin A, and madness. Eyewitness reports are dreadful. Throngs of people reduced to skeletons roamed the countryside and sought help in the towns. The dead were left for a long time, unburied in the hovels or along the streets. The fields were abandoned.

How did the authorities react? On November 29th, 1845 Joseph Stewart, Vice Lieutenant of Ireland (Lord Abercorn, the Lord Lieutenant was absent, just at the right moment), wrote from Port Stewart to Sir Fremantle (University of Nottingham, Department of Manuscripts, NeC, 9196):

> The people here have got the most extravagant notion of the relief they are to get and the priests are determined to magnify the likelihood of famine as much as possible.

A rigid, thoughtless, doctrinaire application of *laissez-faire* principles, a widespread notion that, if he got poor relief, the "unaspiring Irishman" would not work, all fuelled disaster.

Several years after the catastrophe, the *Fraser's Magazine* of September 1868 published an article by a Mr Greg (quoted with approval by Charles Darwin in his book *The Descent of Man* in 1871), in which these instructive remarks could be read (italics in the original):

The careless, squalid, unaspiring Irishman multiplies like rabbits; the frugal, foreseeing, self-respecting, ambitious Scot, stern in his morality, spiritual in his faith, sagacious and disciplined in his intelligence, passes his best years in struggle and in celibacy, marries late and leaves few behind him. Given a land originally peopled by a thousand Saxons and a thousand Celts — and in a dozen generations five-sixths of the population would be Celts, and five-sixths of the property, of the power, of the intellect, would belong to the one-sixth of Saxons that remained. In the eternal "struggle for existence," it would be the inferior and *less* favoured race that had prevailed — and prevailed by virtue not of its good qualities but of its faults.

That the Scot is often, an anglicized Celt just like the Irishman does not seem to have occurred to Mr Greg. Nevertheless, there is little ground to doubt that jokes in poor taste like these were a faithful expression of the current opinion entertained about Irishmen in nineteenth century Britain, and probably much later as well.

While a catastrophe of such magnitude called for immediate extraordinary assistance, very little was done: some public works for people who often no longer had enough strength to perform work of any kind, importation of cereals and their sale at low prices to people who could not afford to pay a farthing for them.

When free soup kitchens were opened in 1847, the third year of the Famine, for a great many people it was too late. Some Anglican clergymen offered soup to the starving Catholics on condition that they became converted to Anglicanism: the initiative met with very little success, and was ironically dubbed "souperism."

Not all Protestant ministers were so callous. One of them wrote an open letter to the Prime Minister, Lord John Russell in these terms (quoted by Kee):

My Lord, I have a right to speak for I am a Minister of God. Let me then importune and implore you, my Lord, to stand in the breach between the living and the dead. (.......) Tell the assembled Parliament that the people must not anymore be left to die.

But in May 1849, in the worst year of the Famine, Lord Russell, speaking in Parliament, had merely this to say (quoted by Kee):

I do not think any effort of this house would, in the present unfortunate state of Ireland, be capable of preventing the dreadful scenes of suffering and death that are now occurring in Ireland. I distinctly repeat that I do not believe it is in the power of this House to do so (.......) I do not feel justified in asking the House for an additional advance of £100,000 which at least would be necessary if the House should say there would be no possible cause of starvation in Ireland.

A similar line of inaction was taken by Sir Charles Wood, Chancellor of the Exchequer, who wrote to an Irish landlord:

I am not at all appalled by your tenantry going. That seems to me a necessary process (.....). We must not complain of what we really want to obtain.

And in fact what "they" wanted to obtain was to squeeze out "excess manpower."
Charles Edward Trevelyan, Head of the Treasury, went on record for a chilling statement like this:

It is my opinion that too much has been done for the people. (.......) Ireland must be left to the operation of natural causes.

And, about his decision to bring the already very scanty relief operations to a close in the summer of 1846, he wrote:

The only way to prevent people from becoming habitually dependent on government is to bring operations to a close. The uncertainty about the new crop only makes it more necessary (.......) Whatever may be done hereafter, these things should be stopped now, or you run the risk of paralysing all private enterprise and having this country on you for an indefinite number of years.

Now these utterances would not pass unnoticed to Irish political leaders, intellectuals and exiles. As Kee (from whom the above quotations are taken) shrewdly remarks,

the whole argument behind the Act of Union (.....) was that the two countries were thenceforth indissolubly bound together for better or worse for all eternity. But if this were not to be so, if Ireland was to be thought of as irritatingly "on" Britain for an indefinite number of years, the Irish might be forgiven for drawing certain political conclusions for the future.

After all, the two countries had been politically and culturally linked for so many centuries, there had been intermingling and intermarriages at all levels between the British and Irish (both of them already a mixture of many European peoples), and even some members of the House of Lords had Irish ancestry. Both Britain and Ireland had Protestant and Catholic populations, albeit in quite different proportions; and confessional toleration, though not always an easy one, had been achieved, at least officially. The Famine, however, and especially the way the disaster was handled by the government, undermined what prospects there might have been for a continuing Union.

But time to think about that would come later. In the meantime, distress was so great as to leave no space but for thoughts of immediate survival. Workhouse inmates soared to 417,000 in 1847 and to 932,000 in 1849, while in 1852 were still 505,000. The number of workhouses grew from 98 in 1844 to 163 in 1848.

The registers of these workhouses are very instructive. Even in Ulster, in spite of the strong Protestant presence, inmates were mostly Catholic who were harder hit by the calamity. This is suggestive of the role exercised by the confessional allegiance in shaping the social stratification. Not long before the Famine, in 1834, the Catholics were the majority (62.5%) even in Ulster. In 1921, at the time of partition, they were decidedly a minority (43%). Evidently the Famine played havoc among them: they starved and emigrated in far greater numbers than the Protestants. Without the Famine, Unionists would have had far greater difficulties in partitioning Ireland.

What follows is an abstract from the admissions register of the Enniskillen workhouse, Co. Fermanagh, reprinted in an educational package published by the PRONI (BG.14/G/1):

1401 / Wilson Biddy / female / aged 16 / single / servant out of employment / R.C. [i.e. Roman Catholic] / not disabled / very dirty / from Ballycassidy, Mossfield / admitted Dec. 2 1846

1402 / McCarth Hugh / male / aged 24 / widower / no employment / R.C. / in Fever Hospital / dirty / from Enniskillen / admitted Dec 3 1846 / left Jan. 7 1847

1403 / Maguire John / male / aged 36 / married / no employment / R.C. / sickly and lame / Siragh Maguire wife / 3 children / very dirty / from Enniskillen / admitted Dec. 3 1846 / left 18 June 1847

1404 / Maguire Siragh / female / aged 40 / married / no employment / R.C. / sickly and cripple / very dirty / from Enniskillen / admitted Dec. 3 1846 / left 18 June 1847

1405 / Maguire Dennis / male / aged 15 / single / no employment / R.C. / not disabled / very dirty / from Enniskillen / admitted Dec. 3 1846 / died 11 June 1847

1406 / Maguire Patt / male / aged 8 / child / R.C. / not disabled / very dirty / from Enniskillen / admitted Dec. 3 1846 / left 18 June 1847

1407 / Maguire Catherine / female / aged 4 / R.C. / not disabled / very dirty / from Enniskillen / admitted Dec. 3 1846 / died January 28 1847

1408 / Mc Cabe James / male / aged 16 / single / begging / C. of E. [i.e. Church of England] / not disabled / very dirty / from Ely Fartha / admitted 3 Dec. 1846 / left Dec. 7 1846

1409 / Kerr John / male / aged 9 / orphan / no employment / C. of E. / not disabled / very clean / from Florence Court / admitted Dec. 3 1846 / left Dec. 4 1846

1410 / McManus Catherine / female / aged 40 / widow / no employment / R.C. / not disabled / dirty / from Union at large / admitted Dec. 3 1846

1411 / Donoghy Mary / female / aged 14 / single / no employment / R. C. / in Fever Hospital / dirty / from Union at large / admitted Dec. 4 1846 / left Dec. 28 1846

The minute book of the poor guardians of Lurgan, Co. Armagh, records, on the 6th of February 1847, a discussion concerning the exceedingly high mortality among the inmates. Within five weeks, from the week ending January 2nd 1846 to January 30th 1847, deaths had soared from 18 to 68 per week. When the Poor Law Commissioners asked for the reasons (PRONI BG.22/A/5/451), the workhouse medical officer explained that

> the great majority of admissions are, when brought into the house, at the point of death, in a moribund state. Many have been known to die on the road, and others on being raised from their beds to come to the workhouse have died before they could be put into the cart, and numbers have died in less than 24 hours subsequent to their admission. Therefore mortality in the workhouse is much greater than under ordinary circumstances, and it is a well-known fact that many dying persons are sent for admission merely that coffins may be obtained for them at expense of the Union (.....).

Overcrowding in the workhouse contributed to the rise in mortality: the document hints that it was impossible to "keep the bedding dry," so that sleeping in "damp beds" caused fevers and bowel complaints, often of a fatal kind.

The staff of workhouses were not always above allegations of corruption, as appears from the minute book of the Clogher Board of Guardians, Co. Tyrone, on 1st May 1849, in which several inmates complain about the porridge which should have been given to them being sold outside, while that served in the workhouse was sometimes "very thin" (PRONI BG.9/A/74).

Keeping discipline in a workhouse was not easy. It appears to have been particularly difficult at Newry. The main reasons may have been overcrowding and the lack of management coordination. Punishments were harsh and remained so after the Famine. From the relevant offences and punishment book we learn that between 1851 and 1855 an inmate was transported for seven years for stealing books and other articles, another was flogged for scratching fresh putty from the window panes, another suffered the same punishment for trying to escape by climbing over the graveyard wall while wearing the Union clothes, one more was flogged for stealing some onions (PRONI BG.24/GJ/1). The Board of Guardians of the same Union, one of the largest in Ulster (230 square miles), passed, on 24th February 1849, a resolution objecting to a rate aimed at providing external aid (PRONI BG.24/A/7/282).

All contemporary sources agree that it was a famine in the midst of plenty. One typical example is a letter from the Rt. Hon. George Dawson, of Castledawson, Co. Derry, to Sir Thomas F. Fremantle, chairman of the Board of Customs, London, on January 17th, 1847 (PRONI T.2603/1):

> There is no want of food; but it is at such a price as to make it totally impossible for a poor man to support his family with the wages he receives. I do not exaggerate when I tell you that from the moment I open my hall door in the morning until dark, I have a crowd of women and children crying out for something to save them from starving. (.....) So great is their distress that they actually faint on getting food into their stomachs. (.....) We are also visited by hordes of wandering poor, who come from the mountains or other districts less favoured by a resident gentry; and worst of all, death is dealing severely and consigning many to an untimely tomb. This week six or seven of the old and infirm have died in my little village, not from want of food, but from the consequence of privation and the total change which has taken place in the habits of the people.

Export of produce from Ireland continued unabated. The island had become an agricultural reserve for Britain, then in the full bloom of the first Industrial Revolution. The flagrant omission of relief to people in immediate danger of death, however, must not lead one to forget the remote cause of the Great Famine in particular, as well as of many other similar calamities which had struck Ireland before: the extreme destitution of a large section the population that had given rise to the tragic dependence upon the potato.

The potato blight destroyed the harvest throughout the British Isles. A tragedy similar to that of Ireland took place on the Highlands of Scotland, causing starvation and emigration, but casualties were in the region of the thousands, not millions. In most of Britain the diet in the countryside was usually richer and more varied, so the worst consequences were avoided. What failed entirely, instead, was the redistribution of other food resources, which were plentiful in Ireland, as in the rest of the British Isles.

The surprising statement by Cullen that famine in Ireland "was caused not simply by the fact that the Irish harvest had failed but that harvests had failed elsewhere as well," certainly cannot be applied to the Great Famine, and is also of dubious value for many other famines in the island.

The Great Famine triggered two sets of consequences: in the socio-economic domain and in the political one. The migratory wave unleashed by the Famine was such as to prevent the population from reaching the historical peak of 1845 anymore. Emigration, however, had uneven effects: the cottier class, more severely hit by mortality and emigration, was wholly wiped out. The tenants, who suffered less harshly, were able to till more land, as did many smallholders.

To the landowner class, the Famine also brought far-reaching changes: many of them went bankrupt, deprived suddenly of their cheap manpower, and due to the repeal of the Corn Laws in 1846 during the Famine and partly as a consequence of it. For them this meant the end of their economically sheltered condition. Foreign cereals could thus reach the British market, in open competition with those grown in Ireland. While agricultural prices dropped, it became more and more convenient to replace tillage with pasture, as the produce of Irish pastoral farming was handsomely paid in Britain.

In 1849, the Encumbered Estates Act eased the sale of properties encumbered by debts and mortgages. About a quarter of the land changed hands between 1849 and 1850. The new landlords were even greedier than the previous ones, more resolute to squeezing quick returns out of their investments. Mass evictions to make room for pastoral farming became an even more common occurrence than heretofore. Already in 1850, the Tenant Right League was formed, in order to fight the evictions.

After the Famine, cities grew, not so much as industrial centres (with the exception of Belfast and the surrounding localities), but rather as commercial nodes due to the swift development of the railways, whose expansion quickened in 1844-45, coinciding with the second and greater wave of the "railway mania" in Britain. Just during the Great Famine lines grew in length by 558%, from 104 to 685 kilometres. The network grew further from 1,392 kilometres in 1854 to 3,813 in 1880. But the cities, in spite of their relative prosperity, or rather just because of it, became focuses of opposition to British domination. The growing independence movements were supported mainly by the rising urban middle classes. Ulster, however, was rather conspicuous for its absence from these developments.

A study of the consequences of the Great Famine calls for a shift in perspective yet further afield, not only on Ireland as a whole, but upon the Irish diaspora, especially to the United States: the population of Irish origin living overseas might be about 20 million, four times as much as the present population of both Irelands together. But, according to a different estimate, in the United States alone, the percentage of Americans with some Irish ancestry is 19% of the total, which would put the number of Irish-Americans to about 50 million.

Leaving aside the Cromwellian deportations to the West Indies, and without reckoning the "wild geese" and the transportations to Australia, the main migrations from Ireland in the modern age involved about 250,000, or perhaps more, Ulster Protestants, mostly Presbyterians, who, as we have seen, moved to North America during the eighteenth century; and 4 million from all Ireland, nearly all Catholics, who settled mostly in the United States.

Although Irish Catholics flocked to the New World by the millions after the Great Famine, they were far less successful than Protestant Ulstermen who had migrated in the previous century. The Protestant migratory wave found in the host country a popolation still small and with a very fluid social structure, so that the immigrants and their offspring had more time to multiply and rise in the social ladder. Many of them, moreover, possessed some little capital to start with. Nor is to be forgotten that Ulster Protestants entered a socio-cultural environment particulary congenial to them. Catholics, instead, found themselves in a Puritan society and, poor as they were, had to start from scratch. Not so massive, yet enough to affect the ethnic character of whole urban districts, was the Irish immigration to Britain, Canada, Australia and New Zealand.

Under the economic viewpoint, emigration became a huge safety valve: it gave those who stayed behind a higher standard of living, both because the same resources were shared among smaller numbers, and thanks to the flow of remittances. As soon as a downswing in the business cycle jammed that safety valve, a sharpening of conflicts was therefore to be expected. It is

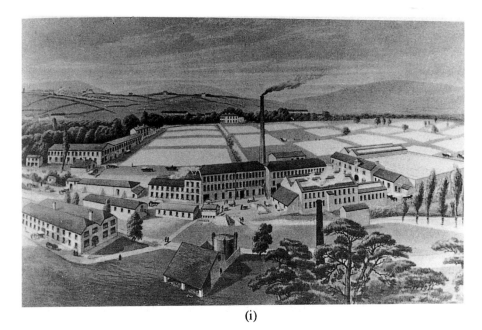

(i)

(ii)

Fig. 3.3 — Two Irish villages in the second half of the nineteenth century which looked totally different: (i) Annvale, Co. Armagh, in Ulster, and (ii) Tullig, Co. Kerry, in Munster. They are good examples of the contrast between the comparative prosperity of the better accessible and more industrialized areas and the de-population of peripheral areas in the aftermath of the Great Famine (by kind permission of Oxford University Press).

precisely what happened with the slump of the late nineteenth century, which brought about a sharpening of agrarian struggles. The dichotomy between the more peripheral areas of the South-West of Ireland, and the more nodal ones, favoured by better accessibility (and therefore preferred by the Protestant colonists), of the North-East, is evident from the most straightforward observation of the landscape (Fig. 3.3).

From the political viewpoint, the large Irish communities in the cities of the Atlantic seaboard of the United States rose rapidly in importance for their mother country. They remained for a long time more Irish than that American, though they were obliged to give up Gaelic in order to achieve a modicum of integration within American society. The discrimination American Irish initially suffered from the Protestants doubtless contributed to making the process of integration longer and more painful and kept alive their hostility against the Protestant domination they had experienced at home, and which they regarded as the very root of their misfortunes. Though still poor by American standards, the immigrant Irish were soon far richer than those who had stayed at home. These New World communities were ready to support any Irish independence movement with the slightest hope of success.

In the United States it was far easier to get firearms. Many thousands of Irishmen took part, on both sides, in the Civil War (1861-65), and thus received practical military training. Frontier incidents with the British in Canada (which thereby hastened to establish its own Federation in 1867), and the beginning of the armed nationalist activity in Ireland were the result. For the first time in history, the Irish had what the strategists call a logistic base, or a "sanctuary," from which to operate: a source of money and weapons out of immediate reach of the enemy.

The tensions in Australia are less known and not so important for the unfolding of the struggle, but nevertheless still noteworthy. Under the Southern Cross, Irish deportees and exiles, and their offspring, contributed to the development of a lively anti-British radicalism. Some of them also formed outlaw gangs of more than local renown.

A more constructive action was that carried out by Irish reformers like Fergus O'Connor, who led the "Chartist" movement in Britain, demanding the universal franchise.

The two main stages of the next act in the drama were New York and Dublin, where, already before the Civil War, in 1858, the Fenian Brotherhood, later renamed Clan-na-Gael, and the Irish Revolutionary Brotherhood (IRB) were founded. A military organization which took the name of Irish Republican Army (IRA) was then formed.

According to a revisionist thesis (Cullen, Johnson), though the Great Famine looks like a "point of no return," great changes were already taking place before the Famine: a decline in fertility, a raise in the age of matrimony, considerable emigration. Just after the Famine the worst affected areas often returned to conditions similar to those before the Famine and underwent no appreciable population decline until the 1870s. The "modernization" of rural life, which tied emigration to changes in the family structure, in agricultural practices (consolidation and rationalization of farms), in population numbers (i.e. a sharp decline; this calls to mind Sir Charles Wood, that voice of officialdom quoted above: "I am not at all appalled by your tenantry going") was more important as a factor of change. In other words the Famine quickened the tempo of change, but did not trigger it.

This point of view overlooks a basic fact: the brutality of the Famine was such that it could not bring about a mere quickening of existing trends, that is to say a purely quantitative change. Instead, it was a definite jump in quality: a *disintegrative innovation*, felt as such by a great many contemporaries, especially the victims. Things could not go on as before longer, and the rise of the Fenian movement is evidence of that. Situations of sharp conflict and social stress do lend themselves very ill to a purely economic approach: social and political implications cannot absolutely be left out of the picture. Even from a strictly economic point of view, on the other hand, the catastrophe was a decisive structural fracture, as recently shown by O'Rourke, among others.

Revisionists, such as Winstanley, who have attempted to appraise the political consequences of the disaster, deny the importance of the Famine in the development of the Irish nationalist movement. They contend that there is no evidence that the sufferers accused the English for their "misfortune." On the contrary, on the basis of the oral tradition, it seems there had been rather a fatalistic acceptance of what appeared as a "natural disaster."

Those who hold these views evidently put great trust in the abilities of objective and rational thinking of a mass of survivors often reduced to little more than walking skeletons. These people, while they tried to escape the hell of the Famine, should have analyzed the situation coolly and perhaps left us written reports: the absurdity of all this does not require demonstration. The ships (popularly called "coffin-ships") which carried the survivors to America, appallingly overcrowded, left a trail of corpses buried at sea. The exiles, moreover, were often illiterate or nearly-illiterate, and had been nurtured in a tradition of resignation supported by religious precepts. And then, during the early years in the United States, they had to tackle pressing survival problems in an entirely new environment, not even a particularly favourable one, at least in the beginning.

It is when survival is assured, that time for reflection is taken, and for telling children the tale of past sufferings. Children, who have a very keen sense of right and wrong, usually feel even stronger indignation than their parents. When some financial means and weapons are at hand, then brooding on retaliation and revenge begins. This is not say that revenge is just or desirable, but merely that it is a common idea of the human mind, and *not* — it should be absolutely clear — of the human mind under extreme stress, but, on the contrary, when it looks back in safety.

The movement for Repeal had very little success before the Famine, though it was led by the "Liberator" Daniel O'Connell. After the Famine, Irish radicals, urban-based on both sides of the Atlantic, perceived the Union as a calamity for Ireland. Catholicism was felt to be part and parcel of the national identity, although the Church did not approve of nationalistic societies and condemned the luckless Fenian uprising of 1867, led by former soldiers of the American Civil War, and the subsequent terrorism of radical groups. The activity of these new radical counter-élites was aimed at a total political devolution of the two islands.

However, the moderates in Ireland were not yet spent. The British imperial élite responded not only with repression, but also with several reforms: attempts at adaptive change which did not, however, reach the hoped-for goal of placating the growing Irish nationalism, fuelled by an impressive cultural awakening in Gaelic sports, language and literature.

The Irish nationalists kept operating on a double track: constitutional opposition, and the revolutionary movement of a pugnacious minority. Land reform, under the pressure of the Land War unleashed by the Land League (founded in 1879), and the drive to Home Rule, developed during the latter part of the nineteenth century. We can only speculate whether the link with Britain might have been preserved had Home Rule been granted without too strong an opposition. The vast majority of Irish Catholics were by no means radical: they sent to Westminster moderate representatives of the Home Rule Party, for whom Irish autonomy within the United Kingdom was the most ambitious goal.

Yet that would have meant an Irish Parliament and an Irish administration, inevitably swayed by the Catholics. The Irish Protestants, particularly those of Ulster, would have none of it. Thus the Unionist ideology was born: a very simple one. It can be summarized in two slogans: "Home Rule is Rome rule" and "Ulster will fight and Ulster will be right," the latter was launched by Sir Randolph Churchill. It found ready support from the British Conservatives, and the constitutional struggle dragged on till the eve of the First World War, when eventually Home Rule for Ireland entered the Statute book, but its application was suspended until the end of the war. It was too late.

Tension ran so high that the British government never dared to introduce conscription into Ireland, notwithstanding the (sometimes desperate) need for military manpower to fight Germany, while both Unionists and Republicans built up paramilitary forces and even imported weapons from Germany a short time before the outbreak of the war.

On Easter Monday 1916, the Dublin Rising broke out. It was a desperate undertaking, and the radical leaders knew it. But in defeat it was successful, due also to the repressive measures adopted by the government, understandably harsh in time of war, yet impolitical. The leaders of the Rising were executed and became martyrs in the eyes of the Irish, heroes like the mythical Cu Chulainn. Internment was adopted against all those who were noted for their Republican ideas, many of them harmless idealists who had not taken part in the Rising. Even lukewarm supporters of independence, when they came out of Frongoch concentration camp and similar places of detention, were turned into fanatics.

In the elections held immediately after the end of the war, the moderates of the Home Rule Party were swept away and in most of Ireland, except Ulster, the radical party Sinn Féin won nearly all the seats. The newly elected Sinn Féiners refused to take up their seats at Westminster and formed the new *Dáil Éireann* ("Assembly of Ireland," or "Parliament of Ireland"). The War of Independence had begun. To win in the field, the British would have needed a new Cromwell, but times had changed and a bloody conquest of Ireland was not to be thought of, though the deeds of the British forces, and especially the infamous Black and Tans and the Auxiliaries sometimes came close to their Cromwellian model, for example, when they set the town of Cork on fire.

After a bloody struggle, the war ended with a compromise. In 1920, the British Parliament passed the Government of Ireland Act, which allowed the partition of Ireland, in order to appease the Unionists. The British government, however, was not particularly anxious to divide Ireland: an outcome which was demanded, even under threat of arms, by the northern Unionists. The latter, on the other hand, did not initially desire the autonomy which was granted to them: all they asked for was to remain politically united with Britain. Eventually they accepted the arrangement because, in the new autonomous statelet, they would have the majority. Thus, the Stormont regime was born: it was to last fifty-one years, from 1921 to 1972.

As is often the case when one tries to satisfy too many irreconcilable interests, as the British had done by accepting the idea of partition, a muddled compromise solution was adopted. It left the way open for a spate of future conflicts.

CHAPTER 4

A LONG-DRAWN STRUGGLE

The stage for the new conflict was set by the Irish Unionists in their internal debate prior to partition. They were split into two groups: the Ulstermen, particularly strong in the six Counties that were to form Northern Ireland, and those of the other Provinces. Their respective viewpoints were swayed by the relative strength of Catholics and Protestants. The latter were a majority only in Ulster, where they formed a nearly complete social pyramid, albeit "heavier" at the top, while in the three southern Provinces they were just a puny minority, mainly at the top of the pyramid (Fig. 4.1).

The difference was also one of economic prospects: the northern Unionists were worried about the damage that might be done to their industrial interests by a government controlled by farming smallholders (a result of land reform) and shopkeepers, as that of Home Rule Ireland was inevitably going to be. The southern Unionists, on the other hand, many of whom occupied highly privileged positions — as bankers, entrepreneurs, members of the professions — in that world of farmers and shopkeepers, had no reason to worry about their economic and social standing.

The Protestant clergy of all denominations sternly opposed Home Rule, while all Catholic bishops supported it. Both in the North and the South, initially Protestants were out for a wholesale defeat of that political innovation, but when Home Rule began to appear more and more likely, the northern Unionists fell back upon partition. Contrary to common belief, the Orange Order did not at first provide an effective ideology and power base to Unionist resistance. The first strong Unionist political organization was the Ulster Loyalist Anti-Repeal Union, founded in 1886. The foundation, in 1893, of the Ulster Defence Union stressed, as the name indicates, the strategic and psychological retreat of northern Unionists: no longer interested in a global conservative defence, which appeared increasingly hopeless, but keen on putting up a less ambitious defence, within the provincial borders.

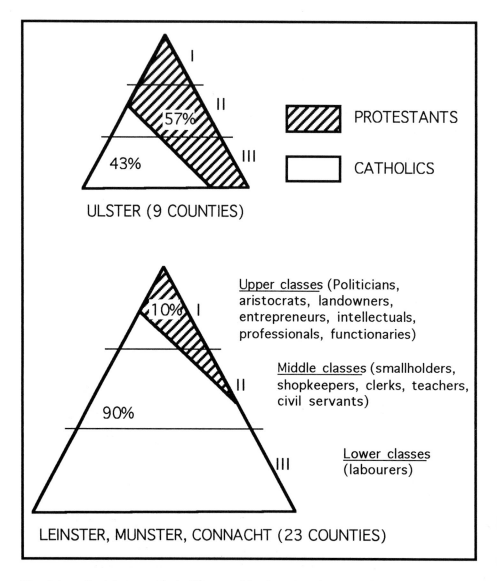

ULSTER (9 COUNTIES)

PROTESTANTS

CATHOLICS

Upper classes (Politicians, aristocrats, landowners, entrepreneurs, intellectuals, professionals, functionaries)

Middle classes (smallholders, shopkeepers, clerks, teachers, civil servants)

Lower classes (labourers)

LEINSTER, MUNSTER, CONNACHT (23 COUNTIES)

Fig. 4.1 — Social pyramids in Ulster and in the other three provinces at the time of partition (1921). In the event, partition did not occur on provincial lines, but Ulster was also partitioned, so that the common usage of the name "Ulster" to indicate Northern Ireland, which has only 6 counties, is utterly wrong.

Instead, southern Unionists, who controlled only three seats in Parliament, did not agree either with the continuous harping of northern Unionists on the theme of the "chosen people threatened by the papists" (one has to bear in mind the strong Calvinistic presence in the North, while the southern Protestant world was dominated by the Anglican Church and was definitely more secularized), or their open propensity to violence. Southern Unionists would have preferred Ulster to remain a part of united Ireland, well knowing that no matter how high they were placed socially, their share of political power would be negligible.

Even the northern Unionists, on the other hand, did not feel secure. Two years after the founding of their paramilitary organization, the Ulster Volunteer Force (UVF), Lord Dunleath wrote to Sir Edward Carson that it was possible to resist the authority of a united Irish Parliament, but a clash with Crown forces was not to be risked, both on moral grounds and insufficient military training (letter dated 15th March 1915, PRONI D.1507/1/1915/7). Quite early on, it became clear, however, that the British government would not resort to military measures to preserve the unity of the country, and there were even doubts about the reliability of army officers had such measures been ordered.

A further split occurred within Ulster Unionism. The leaders of Cos. Cavan, Donegal and Monaghan, as well as some of the other six Counties, favoured a devolution of the entire Province, which would have been economically more accomplished and self-sufficient, but had a mere 57% Protestant majority. The Unionist leaders preferred, however, to give up such an ambitious plan, and be content with six Counties, in which they had, at the time, a more secure 66% Protestant majority. Had the contrary opinion prevailed, a nine-Counties British Ulster would have had a Catholic majority long ago and it might have proved almost impossible to keep it within the United Kingdom.

From the outset, Northern Ireland's existence was haunted by violence. The singular methods used by the regional government to uphold law and order can be gauged from the fact that, in order to put under control one of the most brutal murder gangs which styled itself "Ulster Protestant Association," instead of interning its members by application of the Special Powers Act of 1922 (which was instead rigorously applied against the Catholics), the government enlisted into the police force the whole gang, which could thus pursue its criminal activity under official cover.

Hundreds of people lost their lives in the unrest and terror attacks between 1920 and 1922. Order was brutally restored only by exceptional measures: the Special Powers Act of 1922 allowed the Home Secretary to take any measure and issue any command thought necessary for the preservation of law and order. In pratical terms the police, almost entirely made up of Protestants, had a free hand. The Act was renewed each year, and made permanent in 1933. It remained in force until the introduction of the direct British administration in 1972. On the other hand, a similar piece of legislation is also present in the Statute book of Eire since 1939 to counter IRA activities, and was strengthened in 1987 by an Act for the extradition of terrorists.

This repression of the IRA in the Republic stems from the split which occurred in the Free State between the moderate majority that had accepted the 1920 settlement with Britain which ended the War of Independence, and the radical republicans who recognized only the Republic proclaimed by Patrick Pearse on the stairs of the Dublin Post Office on Easter Monday 1916. The split had caused a civil war throughout the Free State which was longer and bloodier than the War of Independence. In spite of the victory of the moderates in Eire and their outlawing of the IRA, Unionist fears were not allayed: the "southern" State continued to be regarded by the Northern Irish Protestants as a constant threat, no matter what kind of government was in power there.

Besides being a repressive State, Northern Ireland was, from the beginning, a confessional State. A revitalized Orange Order pretty soon took a leading role. The premier Sir James Craig proclaimed:

I am an Orange man first and a member of this Parliament afterwards (.....) We are a Protestant Parliament and a Protestant State.

In full Puritan style, pubs, cinemas, and even children's playgrounds were closed on Sundays. The influence of the Protestant Churches was considerable, particularly in the field of education. No matter how loath they may be to admit it, Protestants in general feel superior to Catholics and ascribe the poverty of the latter to their lack of the virtues of thrift and hard work.

Even nowadays, school separation on confessional grounds helps to perpetuate reciprocal hatred. Both communities disapprove of mixed marriages, though this attitude was stronger until recently. In case of a religiously mixed marriage, the Catholic partner is bound by the Papal decree *Ne temere* to have the children brought up as Catholic. The upshot of all this is that mixed marriages cause distress to the families involved and are a further source of tension.

Recently the term "Protestant" has taken up, at least in some cases, a negative meaning, in the sense of "non-Catholic," of someone who upholds the values of the dominant group, rather than an active member of a Reformed Church. In this perspective one can grasp the meaning of newspaper advertisements such as this, quoted by Heskin:

Wanted — Housekeeper. Protestant (Christian preferred).

According to a current interpretation, the terms "Catholic" and "Protestant" are above all ethnic labels, merely cloaking the age-long antagonism between the Irishmen and the British settlers. All these interpretations doubtless contain more than a grain of truth. There is, moreover, in this regard, an interesting parallel with South Africa, where "Christian" has been, since the colonial age, equivalent to "white," while "kaffer" (a word from the Arabic "kaffir," passed on through Portuguese) means "heathen," and therefore "negro."

In a Northern Irish context, however, the word "Christian" has a peculiar meaning: in the local Protestant parlance, a "Christian," in fact, is he who has experienced the "conversion." This is not, as the term might suggest, a slow personal growth leading to the acknowledgement of truth, but a sudden and entirely subjective inner "enlightenment." It is easy to recognize, in this concept, a strong strain of the most traditional Protestant fundamentalism of pietistic origin.

Pietism, already extant in the thirteenth century in France and the Low Countries, was an important cultural antecedent of the Reformation. It stressed a subjective mystical relationship of the soul with God, resulting in an absolute certitude to achieve salvation irrespective of whatever sin might have been committed, and irrespective of the intercession of the Church or the operation of the Sacraments. These attitudes, in particular attracted the attention of Jean Gerson, Chancellor of the University of Paris and one of the keenest observers of late fourteenth century Europe, in two of his theological treatises (*De diversis diaboli tentationibus* and *De distinctione verarum visionum a falsis*). His masterly treatment of these subjects is one of the major themes of Huizinga's *Herfsttij der Middeleeuwen*.

According to a sample investigation (Belfast Church Survey) of 1983, "conversion," in the sense indicated above, was regarded as necessary to be a "Christian" by 95% of the respondents belonging to the Plymouth Brothers, 92% of the Baptists, 49% of the Presbyterians, 45% of the Methodists, but only by 23% of the sample belonging to the Church of Ireland. This is a significantly low percentage, since the Anglican Communion, in hierarchical structure and theology, is closer to the Catholic Church than most Protestant Churches (or, at least, it was before the controversial ordination of women "priests").

The basis of discrimination, beyond all ethnic implications, is therefore still of a religious nature, and is essentially grounded on fear (Heskin): a fear of the power of the Catholic Church as an organization, although it has many times openly condemned the IRA; of the power of the Roman Church to impose its views on morals and personal behaviour upon its members (Protestants of all hues recoil from the Church's stance on divorce, contraception and abortion, and resent the rearing of children from mixed marriages as Catholics); and finally, a vague, almost primitive uneasiness towards an institution that, to Protestants, appears full of mysticism, symbolism, "clandestine" activity (the Vatican, convents, monasteries, retreats, separate schools) and "unwordly" practices.

Since religion is not so evident as the colour of the skin, several gambits are used in order to discover each other's Church allegiance. The question: "Where did you attend school?" is a very useful one (the segregation of the school system makes the answer immediately revealing), or steering the conversation on the city of Derry (Protestants prefer the plantation name Londonderry, whereas Catholics stick to the pre-plantation name).

Spotting the members of the other group is functional to ethnic social control; for the dominant section of the population in particular, it is essential to discriminate if a firm dominance is to be preserved. In this regard, a minute hold of the territory is very necessary. In the age of the Stormont regime, when they were practically free from British interference, the Protestants went out of their way to achieve that goal: first of all through radical regressive political reforms such as gerrymandering of constituencies, multiple votes to privileged groups of citizens, phasing out of the proportional electoral system that had been introduced by the Parliament of the United Kingdom to provide some protection for the Catholic minority. This was replaced by the Westminster system which enhances majorities.

The two major reforms of this kind were the Local Government (Northern Ireland) Act of 1922, and the Northern Ireland Parliament Electoral Reform Act of 1929. They can be regarded as truly *disintegrative innovations*, that is to say innovations that decrease the integration of the socio-political system and sharpen conflicts by blocking the way to a peaceful defense of opposition interests.

The wholesale Unionist domination in the Stormont Parliament (Photo 4) is evident from a number of facts. During the inter-war period, Northern

Photo 4 — Stormont Building, formerly a sign of Unionist power.

Ireland had only one Premier, James Craig, till his death in 1940. In the Thirties, the region drifted leaderless: Craig's health was so poor that he could not carry out concentrated work for more than one hour, and eventually ended up taking hasty decisions. His ministers were absent, ill, or dying; they seemed unaware of what was going on until they read of it in the newspapers. During his long government, Craig was unable to create the conditions for the survival of Northern Ireland as a political entity. The only clear idea was a negative one: the refusal of unification with Eire and of any cooperation or sharing of power with the large and growing Catholic minority.

The festering social and economic problems were never seriously addressed. The spectre of the republican-nationalist "threat" was functional as a muffling of social tensions within the Protestant population. Only in the eastern part of Northern Ireland, where the Protestant majority was stronger, there arose in the Twenties a lively leftist opposition which was fuelled by Protestant votes, which began to cut into the Unionist majority, until the abolition of the proportional electoral system, besides strengthening the Protestant-Unionist power against the "papists," neutralized the leftist political threat.

No serious complaint was heard from the left-wing Protestants who were disfranchised in this way. They evidently treasured the protection of the Unionist regime more than their own voting rights. For the Protestant poor, in time of economic crisis, the natural outlet of pent-up aggression was to attack the Catholic poor. In July 1935, serious unrest broke out in Belfast between the contiguous ghettoes of Falls (Catholic) and Shankill (Protestant). This resulted in 11 dead, 574 wounded, 367 cases of damage to property, 133 of arson, and more than 300 families, mostly Catholic, made homeless.

After the artificial prosperity during the Second World War, the structural problems of the Northern Irish society resurfaced. But the very poverty of the Catholic population prevented the emergence of an active resistance movement. Moveover, the IRA, outlawed also in Eire, did not have secure sanctuaries in Ireland from which to operate. For this reason, and because of the reaction, for once firm and calm, of the authorities of Northern Ireland, the terror campaign launched in 1956 by the IRA failed. The fact that this campaign was unpopular on both sides of the frontier shows that there was still ample opportunity for reforms meeting Catholic demands.

The most rigid opposition to such a project came, and still comes from the lower class Protestants. Likewise, reforms in South Africa were opposed, on the white side, by the lower classes; while the upper class whites supported, and eventually managed to achieve, the phasing out of apartheid. It stands to reason that discrimination especially benefits the lower social strata of the dominant group: those under more immediate threat by the potential political and economic rise of the lower status groups.

In the Fifties, the stage was set for further clashes. New calls for change came from the rising middle class Catholics who began to get better jobs in the private sector but were still largely politically powerless. In the Sixties, meaningful changes took place in the economy of the British Isles: Eire began to show greater interest in commercial ties with Northern Ireland, the British government refused to keep subsidizing the traditional obsolete industries, especially in Northern Ireland where in 1970 a survey showed that the creation of a new job cost twice as much as in Scotland or Wales.

The prime minister Terence O'Neill, who came to power in 1963, sought to conciliate Catholics, but his initiatives boiled down to a series of symbolic gestures such as the surprise visit of the *Taoiseach* (Irish prime minister) Sean Lemass to Northern Ireland in 1965, and a great many speeches and statements. In this way, he fed both the hopes of the Catholics and the fears of the Unionists. On a practical standpoint, he achieved nothing. His behaviour closely matches that of premier Jan Smuts in South Africa during the Forties, who only succeeded in kindling vain expectations while hardening Nationalist opinion and giving way to the apartheid regime.

In January 1967, the democratic movement of the middle class Catholics found a reference point in the newly founded Northern Ireland Civil Rights Association (NICRA), whose objectives were quite moderate: municipal elections based upon the principle "one man one vote," the end of gerrymandering of constituencies, the setting up of a control structure to prevent discrimination by the authorities, equitable criteria for the allotment of council housing, the repeal of the Special Powers Act and the disbanding of the hated police force, B Specials. There was no question of a united Ireland: NICRA would have been quite satisfied with reform within the United Kingdom.

The new style of pressure used by the Association, based on peaceful marches similar to those of the contemporary civil rights movement of the USA, bewildered the traditional élite of the Protestant group, but certainly not the low class Protestants, from which new élites rapidly emerged, bent on last-ditch resistance and violence. In October 1968, a NICRA march was repeatedly attacked by Protestant mobs and, in Derry, also by the Royal Ulster Constabulary (RUC). Similar incidents took place in the following months, but the Derry outburst had a particularly heavy impact on public opinion because in was broadcast live by the BBC.

This television broadcast of the incidents had a deep, though not lasting, effect on British public opinion. The government of the United Kingdom began to take an interest in Northern Ireland's internal affairs. At the end of 1968, O'Neill announced a programme of reforms which largely accepted the Association's requests, but caused a violent reaction within the Unionist Party. In February 1969, elections were held which, O'Neill hoped, would help overcome the resistance within his party, but the results of the polls were disappointing and two months later he resigned the premiership.

The two following premiers, Chichester-Clark and Faulkner, continued the policy of concessions, as did the British after the prorogation of Stormont, but the reforms came too late, and the position worsened steadily. It should have been clear at the time, as it is clear now, that purely adaptive reforms were insufficient to defuse the conflict.

In 1969, the first detachments of British troups arrived in Northern Ireland. In March 1972, under the N.I. (Temporary Provisions) Act, the government of the region was taken over by a Secretary of State. The costitutional position of the region thus became the same as those of Scotland and Wales. This decision was grounded upon parliamentary enquiries which gave a devastating picture of the responsibilities of the Protestant-Unionist regime. The Cameron Commission of 1969 wrote:

> We are satisfied that all these Unionist-controlled councils have used and use their power to make appointments in a way which benefitted Protestants. In the figures available for October 1968, only thirty per cent of Londonderry Corporation's administrative, clerical and technical employees were Catholic. Out of the ten best-paid posts only one was held by a Catholic. In Dungannon Urban District none of the Council's administrative, clerical and technical employees was a Catholic. In County Fermanagh no senior council posts (and relatively few others) were held by Catholics (.....).
>
> It is fair to note that Newry Urban District, which is controlled by non-Unionists, employed very few Protestants. But two wrongs do not make a right; Protestants who are in the minority in the Newry area, by contrast to the other areas we have specified, do not have a serious unemployment problem, and in Newry there are relatively few Protestants, whereas in the other towns Catholics make up a substantial part of the population.

On the other hand, it was also in Derry, like in Newry, that Catholics were decidedly in the majority, but the control of the Derry municipality was secured by Unionists by means of an incredible gerrymandering trick. For instance, in the municipal elections of 1966, 10,274 Unionists voters elected

12 councillors (with an average of 856 electors for each seat), while 20,102 non-Unionists elected 8 councillors (2,513 per seat). Thus each Unionist vote was worth three times a Catholic one. The Unionist electors controlled two wards: North (8 seats) and Waterside (4 seats). Catholics were concentrated in the ghetto made up by the southern ward, to which 8 councillors were allocated even though it had twice as many electors as the other two wards together. It was only in 1973, at last, thanks to a long-overdue reform, that Derry finally passed under the control of its Catholic majority.

The decision of the Northern Ireland government to allow policemen to form their own Orange Order Lodge, whose first meeting was addressed by the minister of Home Affairs, proves irrefutably the sectarian outlook of a constabulary that should have been above such divisions. The RUC had no training in peaceful mob control and were by no means prepared to face peaceful processions. They resorted to baton-charges and water-cannons, which could but fuel the unrest, which the RUC met with increasing violence, opening fire and causing casualties. In what measure the behaviour of the RUC contributed to unleashing the civil war is plainly shown by the Scarman report of 1972.

According to the Scarman Tribunal there were at least six occasions when those responsible for law and order were seriously at fault: (i) a lack of firm direction in handling the disturbances in Derry on the 12th of August, (ii) the decision to put Ulster Special Constabulary (USC) on riot control duty in the streets of Dungannon on the 13th of August without disarming them, (iii) a similar decision in Armagh on the 14th of August, (iv) the use of machine-guns in Belfast on the 14th and 15th of August, (v) the failure to prevent Protestant mobs burning Catholic houses in Belfast from August 14th to 16th, (vi) the failure to take any effective action to restrain or disperse mobs and to protect lives and property on the 15th of August during daylight hours and before the arrival of the British Army.

The famous BBC broadcast of August 1969 showing peaceful marchers attacked by mobs with nailed clubs, while policemen were only conspicuous for their inaction, aroused strong feelings in Eire. The *Taoiseach* Jack Lynch announced the setting up of field hospitals along the border, and expressed a wish to seek immediate talks with the government of the United Kingdom for a review of the costitutional position of Northern Ireland.

This gave the conflict a new international dimension. The response of Protestants extremists was typical. On the night of 14th August 1969 Protestant mobs unleashed a new terrifying pogrom in Belfast, the main targets being the Catholic ghettoes, Falls and Ardoyne. The RUC behaved in days, seven people were killed and 500 homes burnt down. It was at this stage that the government of Northern Ireland sought the intervention of British troups. A "peace line" in corrugated iron was built around the Catholic ghettoes to isolate them.

The disintegration of the Unionist Party, already begun under O'Neill, accelerated. Initiative in the Protestant field was taken up by extremists such as William Craig and the demagogue Calvinist preacher, Ian Paisley, founder and boss of the Presbyterian Free Church (from 1951) and the Unionist Democratic party (from 1971). Elected M.P. in 1970 for the constituency of Antrim North, European M.P. in 1979, member of the consultative Northern Ireland Assembly since 1981, Paisley is a complex personality, not devoid of intelligence, especially if compared with the wholesale mediocrity in the political world of Northern Ireland's Protestants. He has entertained secret contacts with the Social Democratic and Labour Party (SDLP), finding much common ground, though this party is mainly supported by Catholics; moreover, he has refused any link with the Orange Order, of which he is no longer a member since 1962, as he "discovered" that the organization is anti-democratic. Nevertheless, Paisley did neither restrain nor educate his politically immature followers.

In general, Protestants of Northern Ireland never had any experience of the "give and take" of true democracy. This immaturity is apparent, for example, in the popular belief that the problem of law and order can be tackled exclusively by means of repressive measures, and is strengthened by a vicious circle of economic and cultural backwardness and a dreadful narrowness of ideas, both among the Protestant mass and their political leaders.

A survey in the late Sixties showed that 82% of the residents in Shankill had relatives in the same ward and no less than 45% of the men and 60% of the women had spent their whole existence there (Buckland). It is among the inhabitants of urban areas such as these, who normally perceive the rest of the world (including the rest of Belfast) as a blank, that the mobs for anti-Catholic pogroms have been recruited, as well as the killers of Protestant paramilitary groups, and also, occasionally, some of the victims, when Catholics react or retaliate.

It is true, on the other hand, that Catholics living in ghettoes have a confined range of perception too. Both Catholic and Protestant ghetto dwellers suffer severe poverty, which is by itself a strong factor for isolation. Furthermore, the position of the Catholics living in the Belfast ghettoes is made helpless by discrimination, which leaves them little residential choice and a limited range of human contacts.

made helpless by discrimination, which leaves them little residential choice and a limited range of human contacts.

How do Northern Irish Protestants seek to justify or rationalize their behaviour or, at least, what kind of image do they try to show to the outside world? The answer is simple: they do not deem necessary to justify or rationalize anything; they are as far as they can be from worrying about what the outside world may think of them, similar in this to the Afrikaners of the Fifties. But the South African nationalist regime, fully independent and therefore obliged to face world opinion, eventually felt the need to care about the image of the country abroad and, under Prime Minister Verwoerd, began a well organized propaganda campaign. Moreover, the South Africans were disadvantaged because their "underdogs" were blacks, and therefore "entitled" to massive Third World solidarity.

The Protestants of Northern Ireland, often compared to the Afrikaners, are very far from the painful political maturity the latter have mostly been able to achieve. On the contrary, the Northern Irish Protestant has always tended to identify himself with a broad stereotype of British superiority and insularity, largely outmoded in Britain itself.

The only document I could find which in some way resembles an attempt, by prominent Northern Irish Protestants, to present their own image, is a small volume full of self-complacency, under the title *Orangeism: a new historical appreciation* (Dewar et al.), published in 1967 by the Grand Orange Lodge of Ireland immediately before the outbreak of the unrest. There one can read:

> What mattered to the ordinary [Protestant] man was to be able to feel that his own position and living, and those of his family, were secure. He wanted to go to fairs and markets without being cudgelled there, or waylaid on his return, and to use whatever road he wished. When reports of disorder, intimidation, and "agrarian" crime came in from the South, the northern Protestant refused to allow the slightest self-assertion to "the other side", lest the same occur in his own neighbourhood. On the 12th of July and other occasions, he marched with his lodge behind its flag and drums and fifes, wearing his regalia (cockade, ribbons, scarf and sash) and armed with his yeomanry gun, to show his strength were he thought it would do most good. Where you could "walk" you were dominant, and the other things followed.

Of course, in order to be effective, the march of the 12th of July, in memory of the battle of the Boyne, and likewise the other marches of the so called "marching season" (July and August) have to be provocative (Photo 5). The marchers go through the streets inhabited by Catholics making as

Photo 5 — An Orange Order march. The still picture is unfortunately unable to convey the loud noise which is an essential part of the show.

much noise as possible, re-kindling their resentment. They are thus exhibiting a neurotic behaviour unparalleled in South Africa, where white rule had a comparatively far lower profile, and the ruling group sought to keep its fears concealed as far as possible, refraining at least from such grotesque rites, whereas in Northern Ireland they seem to have a reassurance value for a faltering "master race." A tool and a symbol of dominion at the same time, the marches are a primeval assertion of territoriality. They provide a micro-territorial dimension, street by street, to the conflict dynamics.

Like the South African Broederbond, the Orange Order does not accept female membership, and was closely linked to the party in power, at least before the latter disintegrated.

The differences between the two organizations, however, are greater than the similarities. The Orange Order, founded, as we have seen in 1795 was in the beginning a rural organization which later became partly urbanized. It has about 100,000 members (i.e. one Protestant male out of five, children included) and, far from operating secretly, fulfils the specific function of performing crudely intimidatory shows in full daylight.

The Broederbond, founded in 1918 in Johannesburg, has urban origins and later penetrated the countryside (it was born in fact as a response to the difficulties experienced by the Afrikaners in adapting to urban life). Around 1980, there were about 12,000 members (one Afrikaner male in a hundred), and it operated strictly in secret through a large number of frontline organizations. In fact, all that is known about the South African Broederbond has been gleaned through informers who have passed evidence on to the newspapers.

Both organizations are internal élites, but the Broederbond only can be regarded as a hidden élite. Not even the Royal Black Institution, the inner circle of the Orange Order with over 30,000 members, is so secret as to be comparable with the Broederbond. Its main "rite" is a gathering at Scarva, Co. Armagh, on the 13th of July of every year to re-enact the battle of the Boyne, which obviously cannot be re-enacted in its true historical setting, since the Boyne valley belongs nowadays entirely to the Irish Republic.

Catholics too have their own association, the Ancient Order of Hibernians, which harks back to the time of the Catholic uprising of 1641 and to the resistance movements of the Whiteboys and the Ribbonmen. The Association exists under the present name since the 1830s, and holds its marches on the 15th of August (Assumption Day) and sometimes on the 17th of March (St Patrick's Day).

It would be misleading to regard the Protestants as a monolithic block. On the contrary, the Protestant community is, both socially and from the religious viewpoint, far more fragmented than the Catholic, and this contributes to Protestant insecurity.

Politically there are, as pointed out by Todd, two distinct schools of thought: the loyalist and the "British:" broadly speaking, the former dominates the lower classes, while the latter has a stronger influence among the middle and upper classes. Loyalists are tied to the more uncompromising Protestant tradition, and Paisley is their most representative leader. The "British," instead, regard themselves as liberal and tolerant, they do not agree with the sectarian, aggressive Protestants, and maintain that their Unionism is not the result of anti-Catholic prejudices, but of their upholding the British ideals of good government.

It would be highly desirable that there were many more of these "British". A recent book by Boyle and Hadden makes much of them and the role they, hopefully, could play in bringing about a peaceful solution to the conflcit. But Protestant moderates have, unfortunately, little impact in Northern Ireland. Unsupported by Britain on whom they put so great a trust, they were impotent witnesses to the wreckage of the democratically elected executive in May 1974, as we shall see below.

Moreover, from personal impressions gathered during field work in Belfast, I can but conclude that the ideas of some of these "moderates" are hard to distinguish from those of the most militant extremists. A distinguished academic, noted for being an open-minded "liberal," in trying to demonstrate that Northern Ireland will never become a part of a united Ireland, after much sophisticated reasoning on the comparative population growth of Catholics and Protestants, and after trying to show that the former will never outnumber the latter, uttered a most significant sentence: "In any case, they will never have us, because we are *warlike.*" He then went on to expound the martial virtues, victories and military accomplishments of Irish Protestants.

Such reliance on naked force does not require comment.

There are some small comparatively moderate Unionist organizations such as the Independent Orange Order, established in 1903, after a split from the Orange Order. Although also marching on the 12th of July, as the main organization, it is at least officially prepared to regard Catholics as fellow countrymen.

Decidedly conciliatory and pacifist are organizations such as Women Together, founded in 1970, by Catholic and Protestant women; and Peace People, which came into existence in 1976, and is also interconfessional with a mainly female membership. These two movements held demonstrations for peace and sometimes even tried to interpose physically during riots, seeking to put an end to them. Peace People, whose founders were awarded the Nobel Prize for peace, have also campaigned for reforms.

However, these two movements have received scanty attention from conventional politicians. One of the founders of Peace People, Betty Williams, left Northern Ireland in 1982 and emigrated to the USA after having been subjected to systematic discrimination; two hundred job applications of hers were turned down in eighteen months.

Paramilitary loyalist organizations have proved far more vital than the pacifist organizations. These paramilitaries emerged from the disintegration of the Unionist Party and the weakening of the Orange Order.

Besides these groups, there are organized gangs responsible for the grisly assassination of Catholics, pogroms and plunderings, and a preposterous number of pressure groups and puny political parties, often short-lived. A recent study by Dillon has showed that these organizations have been supported by the British intelligence services in their assassination campaigns, originally intended to eliminate IRA supporters, although it seems that the Northern Irish paramilitaries were also rather wont to kill Catholics at random.

The loyalist secret organization TARA is noteworthy as the closest approach to a Broederbond-type of secret society existing in Northern Ireland. Formed in the mid-Sixties as a pressure group, it took up a paramilitary structure in 1969 at the beginning of the civil unrest (rather euphemistically dubbed "the Troubles"). It went so far as to campaign for outlawing the Catholic Church and closing down all Catholic schools: for all practical purposes, a return to the Penal Laws, extinct more than two centuries ago. This organization is regarded by some as the "hard core" of Protestant resistance.

Also noteworthy is the Ulster Defence Association (UDA), formed in 1971 to thwart the attempt of "the enemies of Faith and Liberty" to destroy Northern Ireland and "enslave the people of God." There is a clear underlying chosen people mentality here, typical of Puritan communities under siege. Similar utterances were, until recently, common catchwords of the Afrikaner conservative extremists in South Africa.

With such a mental attitude, it is not surprising that Protestant violence exploded at the mere idea that "reprobates" alien to the chosen people dared raise their heads, even in a non-violent fashion. Not yet satisfied by the attacks to the Catholic marchers, Protestant extremists unleashed, since March 1969, a spate of bomb attacks, both in Northern Ireland and in Eire.

In October 1969, a member of UVF and of Ian Paisley's Free Presbyterian Church was killed by a bomb he was priming in a power station at Ballyshaven, Co. Donegal.

On the walls of Catholic areas in Belfast one could read at the time this ironic explanation of the monogram IRA: "I ran away." In fact, the Republican paramilitary organization seemed to have been caught napping during the events. Only a vigilante group in Derry, the Derry Citizen's Defence Association, set up in the Bogside Catholic ghetto in July 1969, began building barricades and mounting patrols.

It was only by the end of 1969 that the IRA really awoke, with the split between the "official" wing (OIRA), vaguely marxist in outlook, and the "provisional" wing (PIRA), less familiar with ideology than with bombs and machine-guns. In January of the following year, the political arm of the IRA, the Sinn Féin Party, also split into an Official Sinn Féin and a Provisional Sinn Féin. It was only in February 1971, that the PIRA launched its first terror attacks.

After this renewal of Republican terrorism, directly caused by the obtuse brutality and terrorism of Protestant extremists, the latter had new reasons to feed their own fears and their ill-advised reactions. The *Protestant Telegraph*, a newspaper initially edited by Ian Paisley, voiced Protestant fears in hysterical fashion. Typical, for example, is the following passage, which was quoted by Heskin, and published in 1973, raging against

(.....) the whining multitudes of pestiferous scribbling rodents commonly known as press reporters, newsmen and journalists. This gangrenous population, to be found in every rat hole in Fleet Street, however, are not as perilous as the typhus carriers, i.e. sub-editors and editors. These creatures are mentally flaccid, physically hairless, repulsive and repellent. They usually sport thick-lensed glasses, wear six pairs of ropey sandals, are homosexuals, kiss holy medals or carry secret membership cards of the Communist Party. Most of them are communistoids, without the guts of a red-blooded Communist, or Roman Catholics without the effrontery of a Pope Pius XII. Spineless, brainless mongoloids. But, because of it, as maliciouly perilous as vipers.

Another passage, quoted by Mac Iver, pictures the apocalyptic vision of a "last shore" Protestantism:

The Almighty does not make mistakes; He alone is infallible. Our presence in Ulster is no accident of history (.....) We have an historic and a Divine Commission. We are the defenders of Truth in this Province and in this island.

(.....) Ulster is the last bastion of Evangelical Protestantesim in Western Europe; we must not let drop the torch of Truth at this stage of the eternal conflict between Truth and Evil (.....) We are a special people, not of ourselves, but of our Divine Mission. Ulster arise and acknowledge your God.

According to Ken Heskin, a psychologist of Trinity College, Dublin, with a strong Northern Irish background as a native of Belfast:

> what had been created in Northern Ireland was not a democracy but a "paranocracy" in which the basis of power was the successful appeal to paranoid fears in the Protestant electorate (.....) it was the constant and real fragility of the Northern Irish state, underlined by the continuing claims to sovereignty of successive Dublin governments, which provided the sustaining force of the Northern Irish paranocracy.

Paranoia, as is well known, is characterized by persecution mania, and under this viewpoint the analogy with old South Africa in the times of Strijdom and Verwoerd is by no means a superficial similarity. Northern Irish paranocracy, however, is even more morbidly sensitive to symbolic trappings.

Such are the marches and the obsessive re-enactment, every year, of key episodes of the war between James II and William III. The popular Protestant worship of the memory of the latter king once occasioned the following ludicrous episode, related by Heskin. The Unionist M.P.s at Stormont had a portrait of the "liberator," purchased and exhibited at public expense in the Parliament building. The picture, however, was hurriedly removed when one of the characters depicted in the attitude of hailing, and perhaps even blessing, the Protestant hero, was identified as Pope Alexander VIII.

In fact, the dispute between the Holy See and the French monarchy on the issue of the liberty of the "Gallican" Church had made any kind of Vatican support to James II impossible as the Stuart king had become a protegé of the King of France Louis XIV. The defeat of William III at the battle of the Boyne would probably have made the British Isles a sort of protectorate of the *Roi Soleil*. The European balance of power would have been utterly subverted, with France gaining unchallenged supremacy in the process. This was certainly not what the Vatican and the other Catholic powers wanted. Of course, such historical niceties are very far from the closed and parochial mentality of extremists.

Graffiti is another expression of paranoid symbolism. It serves the purpose of marking territory and scaring away the unwanted Catholics. Extremists have not recoiled from using even human flesh for their scribbling exercises. Repeated attacks were unleashed by Protestant loyalists against the leader of the SDLP, Austin Currie, whose house near Dungannon was hit by gunfire on several occasions. On the 17th of December 1972, while Currie was absent, intruders broke it, brutally assaulted his wife, and scratched with a pointed object the monogram UVF (Ulster Volunteer Force) on her chest.

A great many other examples of paranoid atrocities can be added. Suffice to say that in 1981, an organization named "Silent too long" was set up by the relatives of Catholic victims. This organization maintained at the time that there had been 600 murders of innocent people who had no connection with the IRA, and that the British authorities, since 1972 directly responsible for law and order in the region, had taken no action to unmask and punish the culprits. This must come as no surprise, since covert units of the British military intelligence have traditionally been used in colonial conflict situations, such as in Cyprus, Malaysia, and elsewhere, to make short shrift of opponents and insurgents; the same strategy has been employed in Northern Ireland, where operations of this kind are made easier by the opportunity of employing loyalist hitmen, thereby neatly distancing the government from the responsibility for the killings.

As to the British soldiers, upon their appearance on the scene, in August 1969, they were greeted by the Catholics as deliverers. Their behaviour during the mopping-up operations in the Catholic ghettoes of Belfast and Derry in 1970, however, led to a swift deterioration of their image. Surrounded by feelings of hostility in the militant Catholic areas, the soldiers became pretty soon even more partial than hitherto.

The introduction of internment without trial in 1971 caused intensified unrest. The 30th of January 1972 is remembered as "Bloody Sunday." British soldiers of the First Paratroop Regiment shot dead thirteen people taking part in an unauthorized gathering in Derry. The Civil Rights movement equalled the massacre to that of Sharpeville, South Africa. Though the parliamentary inquest lead by Lord Widgery sought to lay the blame on the victims, international reactions were extremely unfavourable, so that the government of the United Kingdom even made it known that it was considering opening negotiations with the Provisionals. The latter, however, did not trust the offer, and nullified the prospects of their partecipation in such negotiations with a spate of terror attacks, among which were those of "Bloody Friday" on 21st July 1972, which resulted in 11 dead and 130 wounded in 26 distinct explosions in Belfast.

An immediate response to "Bloody Friday" was "Operation Motorman," launched at sunrise on the 31st July 1972: a heavy mopping-up operation carried out by 21,000 British soldiers and 15,000 troopers of the Ulster Defence Regiment and RUC policemen, with the support of armoured cars, in the barricaded no-go areas of Belfast and Derry. Resistance was minimal and only two people were killed, but the whole exercise achieved a result exactly opposite to what was intended, due to the incompetence and brutality with which it was carried out. Incompetence because the vital element of surprise was utterly lacking, and the Provisionals had plenty of time to withdraw from the barricades and go into hiding, only to be back in action a few days later; brutality because the ill treatment inflicted on the population on that occasion, as in many other cases, just swelled the ranks of the Republican guerilla.

Burton, a British sociologist, provides a significant account of the behaviour of British troopers:

A discotheque was interrupted by a foot-patrol who attacked the teenage dancers, putting one boy back into hospital from which he had just been released. He had the stitches from a routine operation on his stomach reopened by the troops. A baker's hand was broken by the soldiers as he went about his delivery round. A store of furniture belonging to homeless, intimidated families was wrecked during a search. Local mill workers were kicked in the genitals as they were searched, twice daily, as they went to and from their workplace. I was hit in the ribs as two Paratroopers asked me if I was in the IRA. I said that I was not and they replied, "Well, fucking well join so that we can shoot you." The friend I was living with was beaten up in the back of a Saracen tank by a Paratroop sergeant. After being interrogated and cleared he was taken back (.....) by the same soldier who apologised, "I'm sorry about that lad on the way down, we do it to everyone. People soon start talking after we soften them up." In addition to these and hosts of other examples, four people were shot dead in heatedly disputed circumstances.

Heskin's analysis, drawing upon a number of highly authoritative sources, is highly illuminating in this regard, and deserves a long quotation:

There is strong evidence that British citizens have been unjustifiably killed and maimed in Ulster by members of the security forces, which is not, in the circumstances, so remarkable. However, the fact that some of these incidents have never been reported in the British media is almost incredible (Mc Cann 1971, *Inside Story* 1972). Mc Cann (1971) notes that "The real sustained and systematic distortion began when British soldiers came onto the streets (.....)." The British publication *Inside Story*, has given examples of unreported maiming and killing by British soldiers. In one instance, an Andersonstown woman was blinded by a rubber bullet fired into her face at point-blank range for ignoring the impolitely expressed instruction of a passing soldier that her window be closed. In another instance, the army, through the honesty of an individual press officer, admitted that they had, in "confusion," shot and killed an unarmed man whom they had previously described as a gunman. The British media declined to report either of these events.

The blame for this state of affairs is variously laid at the feet of the British Army P.R. organisation (*Inside Story* 1972), the Unionist influence in the television media via top executives of BBC Northern Ireland and Ulster Television (*Inside Story* 1972), the British government (*Fortnight* 1973), administrative difficulties and the now repetitive nature of Ulster news (Paxman 1978) and journalists themselves (*Inside Story* 1972, Hoggart 1973). However, one fact, whose shadow falls across all these influences, is that the version of events in any incident involving the British Army emanates primarily and almost invariably from the army itself, "often within twenty minutes of its happening" (Hoggart 1973).

There is a virtual ban on publishing the views of IRA leaders and this ban has been extended at times to include some priests and civil rights leaders (*Inside Story* 1972). The onus has, therefore, been on individual journalists to pursue and examine the veracity of army reports of incidents. With notable exceptions, this responsibilty has been avoided (Hoggart 1973). Bloody Sunday was one of the very few occasions when journalists were on the scene of an incident and not dependent on army accounts of what actually happened (e.g. Winchester 1974).

The accounts of the typical modus operandi of British journalists in Northern Ireland by conscientious British journalists are not flattering. *Inside Story* (1972) paints a picture of newsmen passively waiting in the Europa Hotel, Belfast, for the inevitable telephone call from the British Army P.R. headquarters in Lisburn to supply the basis of their copy describing the latest incident (........).

Both Elliott (1976) and Schlesinger point out that there was a gradual drift away from analysis to a "who, what, where and when" approach which, while enabling the media to avoid the political repercussions of questioning British policy in Ireland, has rendered events in Ireland unintelligible to the British public. As Schlesinger (1978) concludes: "In general, broadcasting presents us with a series of decontextualised reports of violence, and fails to analyse and re-analyse the historical roots of the Irish conflict. Such an approach is largely shared by the rest of the British media, and this cannot but contribute to the dominant public view of Northern Ireland's present troubles as largely incomprehensible and irrational. It is not surprising that many see "terrorism" as the cause of the conflict rather than as one of its symptoms."

It is worth stressing that an open and highly irresponsible call for a decontextualisation of the Irish conflict has also come from certain academic quarters, as we shall see in Chapter 8.

On the other hand, Dillon remarks that the violence of loyalist paramilitaries was "rarely accorded the same column inches in British newspapers" which the IRA violence is regularly getting. Yet there have been more Catholic victims than Protestant, including members of the security forces.

Many cases of torture have been reported, with the exceedingly modern technique of sensory deprivation, through the imposition, for days on end, of a hood which deprives the victim of sight and hearing entirely, applied to suspect members of IRA by British soldiers and the RUC (McGuffin), as well as many further cases of sophisticated tortures and rougher ill-treatment said to have taken place in the detention centre of Castlereagh in Belfast. These facts have caught the attention of Amnesty International, the medical authorities, and the Bennett Commission, appointed in 1978 by the Secretary of State for Northern Ireland.

The report of the commission, however, was the subject of heated argument. It had accepted evidence of ill-treatment inflicted by the police only in 15 cases (0.5%) out of a total of about 3,000 people detained in 1977-78, while such cases seem, on the basis of independent medical evidence, to be far more frequent (Flackes).

British justice and the police have not always been above suspicion, as shown by a case which surfaced in October 1989, when three Irishmen and an English woman (the well known "Guildford Four") were freed from jail. They had been sentenced for life in 1975 as IRA terrorists, on the strength of faked evidence and confessions extracted by means of violence and psychological torture. They were kept in jail, in particularly harsh conditions, although the PIRA unit responsible for the bombing attacks at Guildford and elsewhere had been already apprehended in December 1975 and had made clear that the Guildford Four were not guilty and had nothing to do with the Republican paramilitaries. In spite of that, appeal was refused and the Four were left in jail, while vital evidence showing their innocence was withdrawn by the public prosecution. Only after a passionate national and international campaign, were the innocent freed. Crowning all that, no one of those responsible for so blatantly perverting the course of justice has been punished.

As aptly stressed by Hill and Bennett who provide a harrowing account of these facts, all this could not have happened without a deeply rooted conspiracy at a very high level within the Conservative establishment.

The case of the Guildford Four is not an isolated one. It seems likely that other Irish people, namely the Seven Maguires and the Birmingham Six, are still being unjustly detained, without any redress being in sight. According to Hill and Bennett, the security and judicial system of Britain is liable to cause such things to occur again "any time," having as a contributory factor the deep-seated English prejudice against the Irish. A further point made by Hill and Bennett, which tails perfectly the analyses of the British media quoted above, is the extremely negative role of the popular press, deeply entrenched in sensationalism and gross prejudice. Only the left-wing organizations have been objective and sympathetic towards the Guildford Four and other people in similar conditions.

In connection with the Civil War in Northen Ireland, the British Tory government has been condemned in a ruling by the International Court of Human Rights for the shooting, in 1988, of three unarmed PIRA members in Gibraltar who were preparing a bomb attack. The Court ruled in September 1995, that the action of the British security forces was "unnecessary and in violation of international conventions" (*).

One would logically expect that, in such a tense and repressive situation, the paranoia of one side should be matched by paranoid behaviour on the other side. It seems, in fact, that this is not the case, at least according to the only first-hand account of a social scientist that we we have: Burton, quoted above, who spent eight months, in 1972-73, in a Catholic ghetto under PIRA control. It is admittedly a very old account. If the Republican forces were the bloodthirsty criminals depicted by British mass media, Burton points out, neither himself nor his two English companions could have survived. The only assault he suffered during his field work was from a British patrol: he was beaten without reason and before the troopers tried to ascertain who he was.

Burton found no evidence of the blackmailing activities of which the Provisionals are usually accused, and observed very few traces of psychopathic behaviour. In his opinion, the Provisionals were simply men who defended themselves.

Such defence, one must add, however, has often taken the form of bombing attacks, as usually happens with the weakest side, unable to engage the enemy in open battle and therefore obliged to fight a guerilla warfare. Commercial structures controlled by Protestants were a favourite target, with the evident object to weaken the economic basis of the opponent's power,

---

(*) If I may be allowed a personal note, I will confess that it grieves me deeply to report these things, which throw such a light upon a country I regarded as a beacon of freedom.

and also to spread panic. Here is, for example, a war bulletin in the newspaper of the Provisionals *An Phoblacht* on the 30th August 1984:

> At 11.10 pm, two active service units converged at Fisher's furniture store, William Street, Lurgan. The first ASU, armed with automatic weapons, took up position around the front of the building, while the second group moved into the store and planted three devices. A warning was given, and one hour later the bombs exploded, completely destroying the commercial target. Both units returned safely to their bases.

The Irish Republican Socialist Party (IRSP) has a different character. It was founded in December 1974 following a split within the Official Sinn Féin by extremists who disagreed over the cease fire by OIRA in 1972. In 1975, a bitter struggle raged between IRSP and OIRA; the leader and founder of the new party, Seamus Costello, was killed in Dublin two years later. The military wing of IRSP, the Irish National Liberation Army (INLA), equipped with weapons provided by the former Soviet Union, kept close contacts with international terrorism and was responsible for many attacks in Northern Ireland, Eire, Great Britain, as well as against the British forces deployed in Germany. The Provisionals regard INLA members as "savages" (Flackes). Further feuds have broken out, mostly in Belfast, between the two IRA wings and, in the Protestant camp, between the UDA and the UVF. However, a senior IRA member, talking to Dillon, laid the blame for the feuds between the republican forces on interference by British intelligence.

Violence of paramilitary groups, through the so-called "kangaroo courts," occurs sometimes against their own members, or against inhabitants of areas under their control, for alleged violations of the rules imposed by the paramilitary "authority."

Murray maintains that a high level of violence is related to four factors: (i) the presence of a local Catholic majority, (ii) a history of political violence in the area, (iii) a high level of disaffection of Catholics towards the Protestant society (a rather common occurrence in most of Northern Ireland), (iv) a geographical environment providing the guerilla with both targets and easy shelter.

The main operational areas are: (i) Belfast, especially the predominantly Catholic western area; (ii) Derry, the only place where Catholics were organized for defence since the beginning of "the Troubles" in 1969; (iii) the frontier with Eire, where the rolling terrain covered with vegetation provides a good environment for guerilla warfare.

There are, according to Murray, three strands in the conflict: (i) the guerilla war between the security forces and the Republican paramilitaries, (ii) the campaign of the Provisional IRA against commercial and industrial objectives, (iii) the sectarian conflict between Catholic-Republicans and Protestant-Unionists. In Belfast the second and the third type predominate, in Derry the first one, on the frontier the first and the third. However, even in the "hottest" areas, Catholic support to their paramilitary groups is not wholesale. The Catholic Church has condemned the violence many times and makes strenuous efforts to undermine the influence of the Provisionals and the other Republican paramilitary groups. Perhaps the biggest stumbling block to IRA dominance inside the Catholic ghettoes are the "hoods," gangs of young criminals defying the attempts to keep order both by the police and the IRA itself.

In the second half of the Eighties, the Republican forces abandoned the campaign against shops and industries to concentrate on the security forces and in tit-for-tat attacks against the Unionist murder gangs. In November 1987, however, a dreadful bomb attack took place in Enniskillen, causing eleven casualties. About fifty people were wounded. As a consequence of this outrage, which caused strong feelings on both sides of the border, the Parliament of the Republic passed an act for the extradition of terrorists, thereby laying the foundations for a closer cooperation with the United Kingdom for the repression of terrorism. The atrocities of Protestant extremists, on the other hand, fuelled violence on the opposite side. In 1988 they went so far as to open fire on a crowd during the burial service of IRA members at Milltown graveyard; as a reaction, furious Catholics lynched two British troopers in Belfast right under BBC cameras. Between 1971 and 1988, almost 500 British soldiers lost their lives in the conflict in Northern Ireland.

Though Catholics make up somewhat less than half of the population, 88% of the casualties among civilians are Catholic. A war of this kind, besides the heavy loss of human lives, entails also a high cost in financial terms. This cost should be computed, if reliable data were available, under two distinct items: in terms of property destruction, and of cost of weaponry and resupply. Only a qualitative appraisal can be attempted. Destruction has been heavy on both sides: many families, particularly Catholic, have lost their homes and all their possessions; bomb attacks have played havoc on shops and firms. The Protestants have obviously more to lose in economic terms, but the losses suffered by the Catholic community, which is poorer, are likely to be comparatively heavier.

According to an old but still interesting estimate of the New Ireland Forum, the economic cost of the conflict in 1982 was 1,630 million Irish pounds, of which 1,300 million Irish pounds sustained by Northern Ireland and Great Britain, and 330 million Irish pounds by the Republic. In comparison to the respective Gross National Products, these losses amounted to 23% of the Northern Irish GNP, and 3% of the GNP of the Republic. However, since the Northern economy is supported by the British, Northern Ireland's loss should be compared with that of the whole United Kingdom: in that case, it is but a 0.5% of the GNP.

The impact of the conflict on employment could be serious: the Northern industry is, according to Canning et al., likely to have lost 40,000 jobs between 1971 and 1983, i.e. a quarter of the total since the first year considered. Conflict itself, however, apparently created 24,000 new jobs in security activities, both public and private. Moreover, a substantial fraction of the United Kingdom's budget is being allocated every year to Northern Ireland as a consequence of the conflict, so that the impact of the struggle upon regional employment levels could even be a positive one (Gibson).

Even in that case, however, Northern Ireland would be suffering economic damage, in terms of a loss of her potential for self-reliant development and increased dependence on British financial support. One could take issue against this assessment by pointing out that, in the modern economy, the chances of self-reliant development of such a small and geographically peripheral region would be at best flimsy in any case.

There is a growing male dependency, caused by the rise in female employment paralleled by the collapse of male jobs. Between 1971 and 1986, mostly due to an expanding public sector, female jobs went up by about 40,000 units, a number virtually equalled by the decline of male jobs in industry. In 1986, female employment was 47% of the total and the number of unemployed men was twice that of women.

As to the support to the fighters, the security forces are not a particularly heavy burden to the United Kingdom. Keeping troops in Northern Ireland costs little more than supporting them in Britain and far less than they cost in Germany. The Protestant paramilitary forces are fighting "at home" and need only arms, ammunitions and explosives, which evidently they can get fairly easily, since such paramilitary groups are mostly manned by former soldiers who keep closely in touch with their units and with intelligence covert units. Rather than being a burden, the guerillas seem to provide, for someone, good opportunities for making money. The UDA manages a vast blackmailing business in Protestant areas under its control and has founded several seemingly legal societies. About sixty such societies are listed on the yellow pages of Northern Ireland (Adams).

Financing poses a more complex problem in the case of the Republican forces. The hypothesis of a communist support does not hold as the Irish Communist Party, founded in 1933, has always been extremely weak and has never managed to have a candidate elected either in the *Dáil* or at Stormont. It is but a puny group of a few hundred members, undecided on what to do and largely isolated, though some of its supporters are also in the OIRA. The PIRA, however, is by far the most important wing of the IRA. The only available estimate of its strength dates back to 1970 (Flackes): at that time it seemed that the Provisionals had 1500 members, of whom 800 were in Northern Ireland (600 in Belfast, 100 in Derry, 100 in the rest of the region). However, this data is far from reliable, except as an indication of orders of magnitude.

The weapons used by the Republicans are predominantly Soviet. The USSR did supply the Republican terrorists with arms, not directly, but through the terrorists movements of the Middle East and Gaddafi. Nor did the Soviets provide funds, of which they had not a great deal to spare. Thus these weapons had to be paid for.

Whence did the necessary money come? Probably not, or not mostly, from the Lybian or the Iranian regimes, who were either unable or unwilling to finance the IRA or the OLP to any great extent. On the contrary, terrorism is self-financing. True multinationals of terror have been created. The OLP, for instance, in the mid-Eighties owned about 4,000 million pounds invested all over the world (farms, factories, hotels, real estate, Wall Street shares, gold in Swiss and other banks). The agricultural sector yielded over one third of the aggregate income through large farms in Syria, Guinea, the Somali Republic, and Sudan, and catered alone for the food requirements of the refugee camps, whose control had become a palatable prey to the warring islamic factions. Later on, however, the OLP began to feel the pinch of a severe financial crisis, which has been one of the factors underpinning the agreement with the Israeli government.

The IRA does not reach the size of the OLP, but seems to operate in likewise fashion since, after the murder of Lord Mountbatten (killed by a bomb on his yacht near Mullaghmore, off the Sligo coast in 1979), the FBI disclosed that funds collected in USA by the Irish Northern Committee (NORAID) were used to feed terrorist activities. In consequence, donations from Irish-Americans to NORAID dropped dramatically, and the IRA had to resort massively to self-financing methods.

Such self-financing is probably made up, above all, of tax dodging in the real estate sector, as well as in rackets in the sub-contracting business and faked tax exemption certificates. Having no particularly efficient racket organizations of its own, or perhaps being unwilling to attract too much resentment from the people — as will be remembered, Burton, in 1972-73, found no significant traces of extortionate behaviour, but at that time aid was forthcoming from the United States and the struggle was still basically aimed at self-defence — it seems that the IRA allows Protestant paramilitary groups to penetrate their territory, in exchange for a handsome share. According to Adams, regular meetings are held in a pub in Shankill Road, between Catholic and Protestant terrorists, to share out the booty. After wiping out municipal buses from the streets, the IRA have taken up control of the taxi services in Catholic areas, from which they get substantial profits. IRA men are also accused of being involved in drug smuggling, a charge hotly rejected by the organization.

Self-financing has sometimes been achieved by fanciful methods. The owner of a farm across the border with Eire, in the vicinity of Crossmaglen, took advantage of his peculiar location to speculate on the EEC subsidy for the import of pigs by the United Kingdom, by sending his animals north, legally, through the border, to get the subsidy, and taking them back south within his own enclosure. The same pigs may thus have travelled back and forth hundreds of times. It is estimated that this individual, who has never been caught red-handed, gave the IRA contributions to the tune of a million pounds. A similar trick is used with wheat, whose export is subsidised by EEC when it takes place in the opposite direction, so that, in this case, loads travel legally from Northern Ireland to the Republic with the attendant right to the subsidy, and are then smuggled back to Northern Ireland to begin all over again.

Through these and other no less crafty methods, the IRA gathered enough resources to purchase arms and explosives, usually shipped by means of small cargoes or trawlers, often from Lybia. They were unloaded in some minor port in Eire, and from there the material was carried, in small amounts at a time, to IRA bases, either in the Republic or in the North. The Coast Guard of Eire has already seized several freights. In November 1987, the French Coast Guard seized a ship carrying arms for the Republican forces. The cargo "Eksund," registered in Panama, built in the Netherlands, with an Irish crew, was carrying two hundred tonnes of Soviet weapons, among which, besides AK-47 sub-machine guns, mortars and RPG-9 rocket launchers, all of them regularly used by the Provisionals, there were also deadly land-to-air SAM-7 missiles capable of shooting down planes and helicopters.

In order to broaden their terrorist network, IRA men have established contacts with the African National Congress. According to the International Freedom Forum (a group of independent political analysts based in the United States), the IRA and the ANC planned together the murder of Mrs Thatcher during a visit by the Iron Lady to Zimbabwe in March 1989. Nicholas Mullen, of the IRA, contacted leaders of the *Umkhonto We Sizwe*, the military wing of the ANC and, in October 1988, purchased a house in Harare, about one kilometre from the route Mrs Thatcher was expected to follow.

On the other hand, the government of the South African Republic seems to have supplied the Northern Irish loyalists with arms getting from them, in exchange, parts of the Blowpipe secret missile. After that, in May 1988, the South African Premier of the day, P.W. Botha, was obliged to offer official apologies to the British government. However, three members of the South African Embassy in London were expelled from the United Kingdom.

The British authorities acted, historically, with great resolution in the repression of Catholic paramilitary activities, but often very slowly, perhaps hesitatingly in the quest for a political solution — or maybe they have followed a deliberate strategy of attrition, hoping that, if given time, the unrest and the IRA campaign would subside.

After the introduction of direct administration in March 1972, a referendum took place one year later on the constitutional position of Northen Ireland. It yielded a majority favourable to the continuing political link with Britain. In June 1973, an Assembly of 78 members, having merely consultative functions, was elected, by proportional representation, to replace the dissolved Stormont Parliament. Within the Assembly, an Executive was formed, supported by a three-party coalition made up of a moderate fraction of the Unionist Party, the SDLP and the Alliance Party. The coalition had been agreed upon in December 1973 at Sunningdale between the London and Dublin governments and the leaders of the three parties. Direct British administration ended on the last day of 1973, and since January 1st 1974, power was taken up by the new Executive.

There were also some cosmetic changes, such as the appointment as a commander of the RUC of a Catholic, Sir Jamie Flanagan, in 1973. In that year plastic bullets began to be used in mob control, replacing rubber bullets which had proved lethal on more than one occasion. The new bullets, however, until 1982, caused eleven casualties and a great many injuries.

So far, the action of the British government in tackling the new situation had been tolerably successful. It had achieved the important object of bringing into power the moderates from both sides of the Northern Irish political divide. A fatal indecision, however, surfaced immediately thereafter in facing, or rather in *not* facing adequately, Protestant terrorism and extremist Unionist pressures against the Executive. *The Executive — it cannot be overstressed — was the expression of a democratically elected Assembly, and therefore was entitled to the strongest solidarity and support from the London government and the Westminster Parliament*, a Parliament often regarded as one of the main historical beacons of democracy and liberty. Most unfortunately, however, both the London government and the Westminster Parliament alike failed to act as circumstances required. The situation called for resolute action, which was dismally shortcoming.

Protestant extremists had no difficulty in knocking down the Executive. On the 17th of May, they perpetrated a true aggression against the Republic, blasting three car bombs in Dublin and one in Monaghan. The new Northern Ireland Executive was obliged to resign on the 28th of May 1974, by the general strike called for by the Protestant-Unionist extremists of the Ulster Workers Council led by Ian Paisley, William Craig and Harry West. The strikers cut the electric energy supply, while the Unionist paramilitary groups unleashed yet another terror campaign. The demise of the Executive brought about a political paralysis.

The London government did not apply internment without trial, or any other sanction, to the instigators of such a subversion. It merely took up the direct rule of Northern Ireland again, while the Assembly reverted to purely consultative functions.

It was the first case of a successful political strike in the whole history of the British Isles. Neither the British authorities nor the British public seem to have properly taken stock of the unheard of gravity of such an occurrence. *A democratically elected government left at the mercy of such violence and intimidation had never been seen in the British Isles before. It was something utterly unworthy of a democratic country. The upshot of this could only be a positive encouragement to extremism.*

The least extremist, at that time, were the Catholics. Their electoral behaviour, after the disintegration of the Unionist Party, had shown that very clearly. In contrast, the Protestants had become even more extremist. No seat had been gained by Catholic-Republican groups at the first Assembly elections in 1973, while the Protestant-Loyalist extremists, hostile to the Sunningdale agreement, got 27 out of a total of 78 seats. There was evidently a substantial moderate majority, which included nearly the whole

Catholic electorate and a part of the Protestant one, but, shamefully, *this majority was not allowed to form a stable government.* A precious occasion was irretrievably lost, perhaps in the hope of appeasing the loyalist extremists. Not surprisingly, they were by no means appeased, but rather encouraged to unleash further outrages.

Also the return of successive elections (municipal, for the Assembly, for the Westminster Parliament, for the European Parliament) confirmed, at least until 1982, the trends of the 1973 elections: moderation on the Catholic side and intransigence of a strong section of the Protestant loyalists who, favoured by London's indecision, could baffle with impunity the moderates of both communities.

If the policy of previous governments, both Tory and Labour, had been wavering, with the coming to power in 1979 of Margaret Thatcher, the situation worsened further. Mrs Thatcher was concerned initially with "reconciling" Tories and Unionists; the relation between the two parties having become strained as a consequence of the prorogation of the Stormont Parliament by Premier Edward Heath. Under the Thatcher government, a wholesale hardening towards republican political prisoners led, in 1981, to the death of Bobby Sands and nine of his companions in a hunger strike in the H block of Maze prison. Moreover, Mrs Thatcher kept closely in touch with Enoch Powell and, upon a request from him and other Protestant extremists, increased the seats of Northern Ireland at Westminster from twelve to seventeen. The first political elections, in which the region could send to Westminster the increased number of M.P.s, took place in 1983.

The efforts of successive Secretaries of State for Northern Ireland to bring about an adaptive change failed dismally. Sir Humphrey Atkins, in charge between 1979 and 1981, established from January to May 1980, a Constituent Assembly which failed, due to an Official Unionist Party boycott. In 1981, he proposed the setting up of a consultative council which would have been made up of nominated members only, but the proposal came to nothing.

His successor, James Prior, suggested a "rolling devolution," which should have turned the consultative Assembly into a true autonomous Parliament upon request of at least 70% of its members, or 55 out of 78. This qualified majority was thought necessary to prevent the Unionists from taking control alone. The idea was hailed by the Alliance Party as the "last hope" for a solution of the problem, but opposition against it was wholesale from several different quarters: all Unionist factions, the right wing of the British Tories, as well as the SDLP and the Provisional Sinn Féin. The Eire government rejected the proposal too. In 1986, this attempt also ended in deadlock.

Only in 1982, for the first time, the Provisional Sinn Féin, the political arm of the PIRA, got five seats at the Assembly: undoubted evidence of the radicalization of the conflict and the increasing grip of the Provisionals on Catholic ghettoes. In 1984, confronted by the intractable conflict which threatened England's territory, the Thatcher government, which perhaps had hoped that the struggle would eventually die out, abandoned its old policy of denying the Republic any say in the problem and began holding secret meetings with the Eire government.

After fifteen months of talks an agreement was reached on 15th November 1985 between Mrs Thatcher and the *Taoiseach* Garret Fitzgerald, of the moderate party, Fine Gael. With this agreement, on the one hand it was stressed that Northern Ireland would remain a part of the United Kingdom until the regional majority wanted it, and on the other hand the Dublin government obtained a consultative role in Northern Irish affairs through the creation of a council made up of British, Irish and Northern Irish delegates. This council, in which the Eire government can submit "proposals and viewpoints," has a permanent secretariat and holds meetings in an underground bunker at Maryfield, near Belfast. Delegates are obliged to go there by helicopter for fear of terror attacks by loyalist extremists.

When the agreement was made known, PIRA activity decreased, but the Protestant paramilitary groups unleashed a terror offensive. Unrest raged, with bombing incidents and violence against the Catholics and the security forces. Upon instructions from Mrs Thatcher, finally the police took up a less partial attitude and began to tackle Protestant extremism too. The civil war thus became a triangular one. It would be tedious indeed to relate, even in the most cursory way, the main incidents (communal strife, ambushes, mortar attacks on police stations, murders, and so forth). Particularly serious was the uprising of Belfast Catholics in the summer of 1991, on the twentieth anniversary of the introduction of internment without trial.

But the most important operational theatre remains in England. A special department of the IRA is in charge of attacks on the English "enemy," and *the general persuasion in the IRA command is that the English are quite prepared to accept violence and destruction, provided it remains confined to Northern Ireland. Therefore, the only way for the IRA to be taken seriously by the London government is to take war into England itself* (Dillon). Though the British government has always been very keen to reject officially the very idea of talks with the IRA, in fact contacts between the two sides have been covertly going on for a long time.

The British have been extremely shrewd negotiators, making believe at the same time, in their external propaganda campaign through the newspapers and the BBC, that the IRA was "asking for advice" from the British government itself, allegedly to come out of the impasse. Another stratagem of the London government has been to lead the IRA to phase out their campaign, while at the same time raising the stakes in the talks, until the IRA leaders were left stranded, with no leverage of any kind. According to Dillon, this has occured at least once, in 1974-75, and it accounts for the unwillingness of the IRA leadership to make concessions.

One point in the IRA's stance is obviously wrong. They maintain that the chief obstacle to peace is the British military presence: should the British soldiers be withdrawn, peace and a united Ireland could easily be established, as the loyalists are just Irishmen misguided by British propaganda. How can IRA's leaders fail to see that, in the present circumstances, the main stumbling block towards a united Ireland are the Northern Irish Protestant-Unionists? Do they perhaps avoid acknowledging the truth openly, in the hope of defeating the Protestant extremists after a British withdrawal? This last hypothesis seems more likely, though even in this case the IRA's reasoning may be seriously flawed. It is by no means assured that the IRA forces can inflict a decisive defeat upon the loyalist paramilitaries in open confrontation.

If Unionist obstinacy is doubtless the major stumbling block on the way towards a solution of the conflict, two analyses by the *Sunday Times* and by Buckland, lay much of the blame for this endless tragedy on the British government. The Sunday Times Insight Team wrote in 1972 that, confronted by an age-long conflict, the politics in London has remained the same: a persistent quest for "soft options." The Team came to the conclusion that while this attempt to follow the easiest way is pursued, "the agony will continue." Though this analysis is a very old one, there is still a great deal of truth in it.

The feeble efforts to bring about reforms were suited to British, not Irish, conditions. Attempts to integrate the educational system were slow and belated, thus allowing mutual hate to fester in segregated schools. Integrated schools still serve a puny minority of pupils, though it must be admitted that opposition by the Catholic Church has been a major factor in keeping the schools segregated.

Measures intended to regulate the relationships between the two communities at a local level have followed the pattern of the British legislation for the protection of coloured immigrants. Coloured immigrants in Britain are encouraged to preserve their own identities within the framework of a "multicultural" society. They are, however, few in Northern Ireland, due to the precarious economic situation, and do not bring about any significant problem. But in the Northern Irish conditions the measures involve the keeping Catholics and Protestants apart: segregated, fearful and potentially aggressive.

The setting up of an *ad-hoc* department for communal relations contributed to creating the impression that such relations could be isolated entrusting them to specialists, thereby overlooking the fact that they are of vital concern to anyone in charge of the administration in Northern Ireland. This problem, instead, ought to be borne in mind for every decision taken by any government department.

Finally, the reforms in the police forces have failed to establish an atmosphere of mutual trust, while doubts on British justice have been renewed by the freeing, after less than five years in detention, of a British paratrooper who had killed in cold blood two Catholic youths. The murderer had received a life sentence, but the clamouring of the Conservative establishment, both in the press and in Parliament, where more than hundred M.P.s signed a petition in his favour, eventually had him set free. When this became known, in June 1995, unrest broke out once more in Belfast.

After many years of covert dealings between the London government and the IRA, a ceasefire was eventually achieved early in September of 1994, while the long-standing broadcasting ban on the IRA and Sinn Féin had already been lifted by the Irish government in January of the same year.

As expected, the Protestant paramilitaries intensified their assassination campaign in order to undermine the ceasefire. Not even young girls and pregnant women, whose only fault was to be Catholic, have been spared, making patent nonsense of the current Loyalist contention, that their paramilitaries are merely reacting to IRA violence. To show "goodwill" the IRA had to refrain from reacting to the violence of the Unionist paramilitaries for one and a half months.

Eventually, by mid-October, the Combined Loyalist Military Command also agreed to the ceasefire, but only after receiving guarantees that the constitutional position of Northern Ireland within the United Kingdom will remain unchanged. As so often in the past, the British authorities have bowed to the Unionists. Conflict resolution, therefore, is not exactly at hand, at least under the present Tory government. By late October 1994, the London government was satisfied at last that the ceasefire could be regarded as permanent, and reopened the border between the two Irelands.

Propaganda maneuvres, however, had just begun. For instance, late in November 1994, the *Sunday Times* published an article stating that a splinter group of the IRA was threatening to unleash "a new war." According to Sinn Féin, the whole story was but a carefully placed red herring by the British Secret Service to discredit the IRA, presenting it as unable to control its own affiliates.

After much wrangling between the London government and Sinn Féin about the decommissioning of IRA weapons and newspaper leaks of British-Irish political deals, a clear attempt to achieve an agreement between the British and the Irish governments, virtually cutting out the IRA, came to light when, in February 1995, the Premier John Major and the *Taoiseach* Albert Reynolds presented together in Belfast a proposal for self-determination in Northern Ireland.

The proposal is intended only as a point of departure for further discussion and not as a set programme; therefore it is not going to be enforced against the will of the majority. All parties in Northern Ireland were invited to contribute to the debate. The document envisages a new Northern Ireland Assembly, elected by a proportional system, with extensive legislative and executive powers, except those concerning constabulary and security matters. A committe of three "sages" will be elected by the Assembly to monitor its activities. It seems to be but another "rolling devolution": a solution that has been tried before and has failed. Moreover, a new North-South Body made up of representatives from the Dublin *Dáil* and the Northern Ireland Assembly will be formed, but having extremely limited powers. Finally, both the United Kingdom and the Republic of Ireland promise to amend their respective Constitutions: Ireland will abolish any claim to sovereignty over all Ireland, while Britain will suppress any reference to Northern Ireland as an inalienable part of the United Kingdom.

The Dublin government seems therefore to have accepted the "soft options" strategy of the British, according to which the final decision should be left to the inhabitants of Northern Ireland alone. As Major puts it, an agreement will have to be reached by the parties, by the inhabitants of Northern Ireland and be approved by Parliament: it will be a procedure "of threefold consent" (sic).

A procedure of *assured paralysis*, one should rather say. Unionists reacted furiously. "Betrayal" and "sell-out" were the most common expressions used in their comments. Ian Paisley immediately dubbed the agreement "a one-way road to a united Ireland." This is a foretaste of the kind of consent that can be expected.

Are the British serious or do they hope to engineer a remake of 1974-75, when, during secret negotiations, they raised the price more and more and displaced the IRA? Will the ceasefire of the autumn of 1994 prove to be just one more trick to cajole the IRA forces into decommissioning their arms, and so reducing them to irrelevance? Also, what will the British authorities do if a new Northern Irish Executive, supported by the Catholic and Protestant moderates, were to undergo a shattering attack by loyalists through strikes and intimidation as the Executive of 1974? Will they again watch passively?

While the British Conservative government have been adamant, as usual, that no change in the constitutional position of Northern Ireland will be made without the consent of the local majority, i.e. of the Protestant-Unionists, at the same time they have avoided stating what their final aim is, if indeed they have any. Incertitude is thus likely to worsen existing tensions.

There is very little hope that peace will be achieved in this way, and even reason to doubt whether the British Tory government are just buying time in order to leave the hot Northern Irish potato to others. In the realm of politics, helas, the worst suspicions are too often quite justified.

The result of all this is renewed war. On the evening of Friday 9th of February 1996, the IRA announced the end of the ceasefire, as no progress whatsoever had been made in seventeen months of negotiations. The same evening a powerful bomb exploded in the London Docklands, close to the headquarters of the News Corporation: the huge newspaper trust belonging to media tycoon Rupert Murdoch, owner of the main Conservative-oriented dailies, including the *Times* and the *Sun*.

CHAPTER 5

AN IMBALANCED SOCIETY

The last three chapters have been concerned with the historical development of Northern Ireland and have led us to the latest events at the time of writing.

Concentrating on the latest news, however, does not help us a great deal in understanding the situation. To achieve that understanding, we must turn again to the social imbalances underpinning the present conflict: they are the subject of this chapter. The following chapters examine, respectively: planning, housing and the economy; the development of the regional capital, Belfast; the possible interpretations of the conflict process; and finally, the future prospects of Northern Ireland.

Colonial control, plantation of settlers and relationships with the British "mother country" reached a high level of efficiency in North-East Ulster, especially in those parts of Cos. Antrim and Down which gravitate directly to Belfast Lough. The high potential accessibility of this part of the island has been exploited by the introduction of a more dynamic social system, following the plantation.

The replacement of the indigenous subsistence economy by the commercial one, the intensification of transport networks and above all the emergence of the Industrial Revolution have enhanced more and more the locational advantage of this part of the island, thereby increasingly strengthening the spatial imbalances.

These overlap with the social imbalances generated by the colonial domination. In a broad sketch of the *ancien régime* conditions, we have seen in Chapter 3 that, in the countryside, the upper layer of society was made up of landowners. The Anglican confession, supported by the British imperial power, was the reference point for these magnates, as well as for other groups (lower Anglican clergy, farmers, shopkeepers) of English stock, who formed the set of first-class citizens and suffered no political discrimination. The Presbyterians were second-class citizens: they were mostly of Scottish stock, farmers, tenants, middlemen, superintendents, artisans, but also labourers. The majority of labourers, however, were Catholic and, though there were also many tenants of this confession in marginal parts of Ulster, the Catholic population was doubtless made up of third class citizens, due to their depressed economic position and the heavy political discrimination under which Catholics laboured.

In the urban environment, a powerful Presbyterian population gave rise to a flourishing commercial and entrepreneurial class, whilst the Catholics made up, as usual, the most disadvantaged social strata. Among Protestants too there were, however, large disadvantaged classes, so that Protestant ghettoes grew often quite close to the Catholic ghettoes. The great landowners, usually belonging to the Established Church, exercised a heavy tutelage upon the cities, as for instance did the Chichesters on Belfast.

As we have seen, the Industrial Revolution came late to Ireland, with the exception of North-Eastern Ulster, where dependency upon Great Britain, rather than being an obstacle to development, was a stimulating factor. It was, of course, not to the advantage of the native Irish (who were, on the other hand, very far from the activism of a dynamic society, most necessary for economic success in conditions of acute competition), but to the colonists, above all the Scots, less strongly represented in the landowning class and bearers of a tradition of economic dynamism and practical education. One more factor of development was the settlement, towards the end of the seventeenth century, in the valley between Belfast and Lisburn, of a colony of Huguenot refugees, bearers of advanced technologies in linen manufacture, especially bleaching. A significant role was also played, since 1825, by banks, set up with capital from Liverpool or from Ulster itself: financial resources were thus conveyed from agriculture and commerce into industry.

From a commercial viewpoint, the introduction of free trade with Great Britain opened, in the first half of the nineteenth century, a vast market for Ulster linen. Already in 1800, Irish linen production was more than sufficient for the internal market. In that year 32 million metres of linen textiles were exported, mostly from Ulster, while in the rest of Ireland production was rapidly declining. Not even Dublin, the first city to have a Linen Hall open since 1728, half a century before its Belfast namesake managed to compete successfully with eastern Ulster. In 1850, mechanization had made a powerful headstart in this area: over three quarters of the Irish spinneries were concentrated in Cos. Antrim and Down, while bleaching plants were located mostly in the valleys of the Lagan and the Bann.

The linking of innovations, typical of the Industrial Revolution, led to an overall improvement in the bleaching, spinning and weaving technologies, as well as to a vertical integration of the production. Spatial links were intensified, as the Industrial Revolution integrated eastern Ulster within the British space economy, and estranged the area, even more than it already was, from the rest of Ireland. However, it also caused an increase of territorial imbalances within Ulster, strengthening an inequality between the central and peripheral areas which is still very much in evidence today.

Peripheral and intermediate areas are, on the other hand, those having a stronger Catholic presence, while the central area, gravitating upon Belfast, has a decided Protestant majority (Fig. 5.1). Segregation is such that, usually, in local administrative units Catholics tend to be either a massive majority or a helpless minority. On a small scale, segregation may be particularly strong. Peripheral areas suffer appalling unemployment rates.

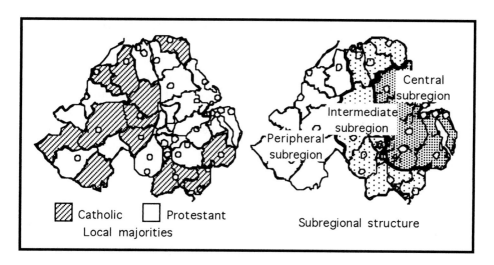

Fig. 5.1 — Relationships between central and peripheral areas compared with local majorities. Although based upon old data, the pattern shown here still holds. If anything, segregation has tended to increase. The maps are based upon a subdivision of present-day Northern Ireland into 67 local administrative units (Source: Hoare).

In my previous book, I had sought to offer an overview of the social and spatial imbalances in Northern Ireland by means of an analysis of data from the 1981 Census, the latest one, at that time, whose returns were entirely published. The statistical technique I used is called principal components analysis. Without going into any technical detail, one can say that this technique allows one to "collapse" together variables whose behaviour and distribution in space are sufficiently similar. The percentage distribution of fifteen variables in the twenty-six administrative districts set up by the 1971 Boundaries Act was taken as a database.

The main component extracted from the data accounted for 39.9% of the total variability of the data (variance), and revealed a close association between the variables indicating faster population increase, a high number of young dependents, bad housing, unemployment, and Catholic population (Fig. 5.2). A further component, accounting for 22.5% of the variance, revealed instead a latent dimension of comparative poverty and industrial employment, with no well defined pattern of religious affiliation. Further components were of decreasing significance and will not be considered here.

The existence of a serious social and spatial imbalance in Northern Ireland is, therefore, quite evident. It would be important at this stage to find out whether such imbalance triggers social pathologies (e.g. poverty, alcoholism, drugs, illegitimate births, criminality), and if all this is linked in some way to the population imbalance, i.e. to a tendency of the socially disadvantaged population to grow more rapidly than the dominant group.

In some cases, it is possible to pinpoint the differences among the different ethnic groups as far as a set of social pathologies are concerned, such as, for instance, criminality. I have been able to do so myself in a study on South Africa. The dearth of data makes such comparisons unfeasible for Northern Ireland. One can only remark that life expectancy in the region as a whole (70.2 years for males and 76.5 for females) is slightly below the average for the United Kingdom (respectively 71.7 and 77.5), which might perhaps be a consequence of the lower living standards in the region. Data from the Republic, on the other hand, are close to the Northern Irish ones, and all these differences are probably too small to draw any conclusion.

It is, however, possible to compare crime rates in the Republic with those of Northern Ireland. The unreliability of this kind of data is well known, as they include only crimes reported to the police. The comparison concerning murder is particularly suggestive: in 1983 there were 0.7 cases for 100,000 inhabitants in Eire, and 5.8 in Northern Ireland. It is more difficult to compare this kind of data for the following years, due to the wide divergence in statistical criteria: in the United Kingdom the number of convictions is given, while in Eire statistics give the total number of murder cases.

It is nevertheless interesting to note that in England and Wales (Scotland cannot be included, again due to discrepancies in data reporting, as there, the number of accused persons is given) the number of murder convictions

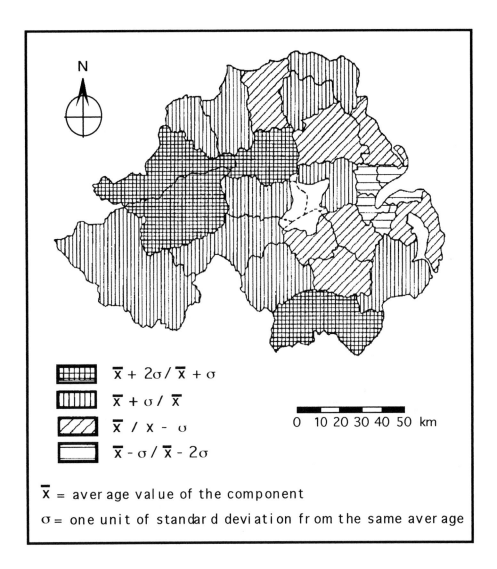

Fig. 5.2 — First component (accounting for 39.9% of total variance) of principal components analysis of economic and social imbalances in Northern Ireland. Explanation in text.

compared with the number of inhabitants, though still far below that of Northern Ireland, has been gradually approaching the Northern Irish level. In fact, while in 1977 the number of murder convictions in Northern Ireland had been nearly seventeen times higher than the corresponding figure for England and Wales (334 for 100,000 inhabitants against 20), in 1987 the figure was only four times higher (160 against 40). This seems to point to a comparative decline of violence in the region.

Leaving violent deaths aside, the mortality pattern does not differ significantly from that typical of developed countries, with heart diseases and cancer as the main causes. In the study on South Africa previously quoted, it had been possible to distinguish clearly the mortality pattern of the white dominant group from those of the other groups, while in Northern Ireland a similar comparison between Protestants and Catholics is impossible due to the lack of separate data for the two groups.

Large numbers of Catholics live in ghettoes and suffer from extreme poverty, while conditions are not so severe for Protestants, of whom even the poorest have been able, while the Stormont regime lasted, to take advantage of anti-Catholic discrimination in terms of jobs and council housing allocation. Even now, due to intimidation by their paramilitary groups, Protestants are able to influence decisions in housing and can still derive benefits from the stronger economic position of the Protestant community as a whole. The marginal position of the low class Protestants arises rather, as in the case of the Afrikaners in the apartheid era, out of an isolationistic attitude from a world perceived as hostile to the "chosen people," and also, more understandably, from poverty.

A good indicator of social pathologies, in particular poverty, could be emigration, if suitable separate historical series for Catholics and Protestants were available. However, this is not the case, and even aggregate data is extremely inaccurate due to the difficulty of controlling movement across the border with Eire. It is plain, however, that the migration balance has always been negative. Since the partition, a net outflow of over 400,000 people has occurred from the region. The excess of emigrants over immigrants was about 92,000 people between 1951 and 1961, 62,500 in the decade of 1961-71, and 153,000 in the difficult decade of 1971-81, a time of unrest and economic crisis.

Between 1961 and 1971 it seems that the negative migration balance was 24,500 for Protestants (an yearly net outflow of 2.6 ‰ of the total Protestant population) and 44,500 for Catholics (8.5 ‰ of the Catholic total). No comparable data by confessional affiliation is available since 1971. Emigration has no doubt a strong conflict dimension: the rigid anti-Catholic discrimination in jobs and housing allocation under the Stormont regime, was precisely aimed at compelling Catholics to move out, with the evident purpose of preserving Protestant dominance. However, it is impossible to say what percentage of Catholic emigration was due to economic reasons and what instead was mainly triggered by discrimination itself.

In order to gauge whether the difference between the population dynamics of the Catholics and that of the Protestants is linked in some way to social pathologies, it would be important to evaluate the difference between the percentages of illegitimate births in the two groups. The only comparison of this kind known to me has been carried out by researchers of the Queen's University of Belfast on a sample of 2988 women. There were 44 unmarried mothers in the sample: 25 Catholic and 19 Protestant (Paul Compton, personal communication of September 27th 1984). In other words, 56.8% of illegitimate births in the sample were to Catholics, even though the latter were only 38.5% of the total. However, the smallness of the sample does not support any firm conclusion.

It is therefore possible, though by no means sufficiently proved, that the fertility differential between Catholics and Protestants is at least partly due to a higher number of illegitimate births among Catholics. It has been suggested that their status of discriminated minority is a direct cause of higher fertility.

According to Compton, however, this hypothesis does not hold because it fails to explain why the Catholics of Eire, who do not suffer of a minority condition and are not currently subject to discrimination, have a fertility index similar to that of Northern Irish Catholics. On the contrary, the comparatively higher fertility of Catholics, both in the North and the South, could rather be accounted for by their obedience to the teaching of their Church against birth control. Moreover, a socioeconomic explanation seeking to link high birthrate with poverty and unemployment, seems to Compton unable to explain why the lower class Northern Protestants, though suffering similar social problems — which he rather uncritically regards as entirely equivalent to those suffered by Catholics — have a low fertility rate. Neither would it be possible to explain the comparatively higher fertility of well-to-do Catholics.

The fertility imbalance between the Catholics and Protestants is evident since the availability of the earliest population data. However, given the low marital rate of Catholics and their age structure considerably altered by emigration, this imbalance did not significantly affect birthrates until the Thirties. Later on, however, the difference came powerfully to the fore. In 1937, the Catholic birthrate was 22% higher, and forty years later 51%

higher than that of the Protestants.

Net migration, i.e. the difference between emigration and immigration, only partially makes good this imbalance. Obviously, great caution is required in the interpretation of the data. In 1981, the percentage of Catholics was ludicrously low: only 28%. That was a time of severe civil unrest; a campaign was waged against the census and a very high percentage (18.5%) of population did not answer the question on religious affiliation, especially in the Catholic areas.

In 1991, there was instead rather an incentive to state the religious affiliation correctly, as religion is being increasingly used in decisions concerning the territorial allocation of resources: the official percentage of Catholics was 38.4%. The problem of estimating the "true percentage" of Catholics kept population geographers, demographers, political scientists, and other professional meddlers very busy. "True estimates" ranged from an "optimistic" (for Unionists) 38.5% to a "pessimistic" (again for Unionists) over 41%. In 1991, "non-statement" was only 7.3%, and higher in Protestant areas. This raises an interesting but difficult question: does it mean that in 1991, Protestants were more underestimated than Catholics, or that Catholics in Protestant areas preferred to avoid answering the question on religion? If the percentage of non-respondents was equally divided between the Catholics and the Protestants, the former would reach 42%.

Be that as it may, the Catholic marital rate rose from under 30% to over 40.2% of the total since 1927: this caused the Catholic birthrate to remain substantially higher at least since the 1951 census, notwithstanding their higher emigration rate. This difference holds even when an allowance is made for the different socioeconomic structure of the Catholic and Protestant sections of the population. In simpler words, the former are poorer in average and conspicuously more represented among the lower classes. Yet low class Catholics have more children than low class Protestants, and the same holds if we compare the fertility rates of the middle and upper classes of the two groups.

No doubt, some population geographers and demographers will be glad to point out the methodological weaknesses in this analysis, especially if their heart lies in the preservation of the political link between Britain and Northen Ireland. It can readily be admitted that far more sophisticated statistical techniques could be brought into play. A "brilliant" analysis of this kind, however, would probably be like using a modern computerized piece of artillery to shoot reject ammunition. As is well known, census data, to put it mildly, sometimes has a low level of reliability: inaccuracies of different kinds, tampering with the returns to suit different vested interests, problems of obtaining information in socially difficult neighbourhoods, all contribute to make census returns somewhat unreliable.

In a highly tense situation, when population numbers are crucial to the outcome of a harsh conflict, the data becomes even more suspicious. It would be highly instructive, especially for supporters of the effectiveness of sophisticated demographic techniques, to watch a census enumerator of Her British Majesty in action in an IRA dominated neighbourhood of Belfast or Derry, bearing in mind that the last census year, 1991, was also an year of considerable Catholic rioting. In addition, complex demographic models are likely to delight the professional tribe but would only confuse the average reader. Finally, it is quite impossible to be fully persuasive, one way or other, in a discussion on such an intrinsically difficult and emotionally charged topic. Time, rather than statistical formulae and computer hardware, is likely to tell a more credible story.

According to Compton, the differences in birthrate can partly be explained by the higher percentage of Protestant married women having a job, who are more likely to practise birth control. Does this mean that such a difference is merely linked to sociocultural factors and not to religious affiliation? As pointed out earlier, however, Protestants in general, and of course, Protestant women, enjoy more job opportunities thanks to the diffused anti-Catholic discrimination, either current discrimination in private employment or in Protestant-controlled local authorities, or the heritage of the past wholesale discrimination from the time of the Stormont regime. Thus, under this viewpoint, socio-pathological conditions come again very much to the fore.

Compton, who denies links between population dynamics, social pathologies and conflict, formulates a simple, perhaps too simplistic model to describe — hopefully not to *explain* — fertility differences in the British Isles. The model assumes two typical demographic regimes: that of the Protestant population with a high marital rate and low birthrate, and that of Irish Catholics with a lower marriage rate and, in obedience to Church dictates, a higher birthrate. The two regimes, according to Compton, just overlap in Northern Ireland. It would be interesting, however, to know what position would be taken, in this model, by the British Catholics.

Models of this kind may yield awkward results. South Africa could be viewed as a space in which a low fertility European regime overlaps a high fertility African one. Likewise, formerly Soviet Central Asia could have a low fertility European (Russian) regime overlapping a high fertility Moslem one. And are we to regard the high fertility of USA blacks as the result of an "African" regime? The best that can be said of models of this kind is that they are tautological, i.e. they merely restate what is already known without providing one jot of explanation.

Socio-cultural factors are definitely important in shaping demographic behaviour. But one cannot ignore the psychological dimension of an oppressive discriminatory situation. This delays modernization in all its aspects, economic, social, psychological and, inevitably, demographic. In the ruling group, instead, privilege leads to individualistic and materialistic behaviour patterns, and birth control is more often practised in order to preserve living standards. On the other hand, it would be futile to deny the paramount importance of the attitude towards religion in shaping fertility regimes.

Evidence of the conscious use of reproduction as an openly declared weapon is not hard to find, as, for instance, in South Africa, the Palestinian Territories and Lebanon. It is difficult to believe that such is not the case in Northern Ireland where the universal franchise has been long in existence, and where everybody knows that every birth is a potential future vote.

The issue of fertility has always been strongly felt in Northern Ireland on both sides of the divide. In a radio interview a few days after his resignation from the premiership (*Belfast Telegraph*, 5th May 1969), Terence O'Neill expressed his opinions on the issue with crystal-clear openness, not disguising, at the same time, his rather glib identification of Protestantism with civilized living and of Catholicism with authoritarianism, typical of Northern Irish Protestant prejudice:

> The basic fear of Protestants in Northern Ireland is that they will be outbred by Roman Catholics. It is as simple as that. It is frightfully hard to explain to a Protestant that if you give Roman Catholics a good house they will live like Protestants, because they will see their neighbours with cars and television sets.
> They will refuse to have eighteen children, but if the Roman Catholic is jobless and lives in a most ghastly hovel, he will rear eighteen children on national assistance.
> It is impossible to explain this to a militant Protestant, because he is so keen to deny civil rights to his Roman Catholic neighbours. He cannot understand, in fact, that if you treat Roman Catholics with due consideration and kindness they will live like Protestants, in spite of the authoritative nature of their Church.

Moreover, Compton's model would benefit from a broader historical perspective. Can we speak at present of a demographic regime of the Northern Irish Protestants because they have kept a given behaviour over some years? What behaviour? For how many years? And what does it mean to say that a given demographic regime is typical of Protestants or

Catholics? Protestants have bred very rapidly in other parts of the world and at other times, for instance in North America until the late 19th century. Catholics in many countries have in these days an extremely low fertility.

Minority status is in itself socially pathological if the minority is not respected and protected. Such a condition is felt by Irishmen as a whole, not only by Northern Irish Catholics. This is clearly expressed by leading historian O'Beirne Ranelagh:

> (.....) over the centuries of increasingly powerful and centralized British government, ruling social and political pressures combined first to make Irish people feel and then to believe that they were inferior. This is one of the worst things that any nation or race can do to another. It results in the most terrible of paradoxes where in practical matters there is a desire equally to welcome and to oppose, thus ensuring that failure accompanies success, and despair and a sense of futility underlie the whole of life. As many Irishmen were government spies, agents and informers as were national heroes; emigration became almost the only way to escape depression. To the present day many Irish writers find it somehow necessary to practise their art away from home.

These masterly words should be borne in mind by anyone seeking to understand Irish history and culture, and the conflict in Northern Ireland in particular. Like Irish writers practising their art "away from home," many leading Irish historians have also "exiled" themselves, either physically or mentally, by accepting and abetting a curiously edulcorated version of Irish history, from which colonialism, devastation, discrimination, humiliation, seem to recede into a dim limbo.

Although anti-Catholic discrimination has not existed in Eire for generations, Irish culture has been by now irreversibly altered, population density has never reached pre-Famine levels, British economic power (albeit to a lesser extent) is still paramount, and the Irish historical record, in spite of any possible "revisionistic" fashion, is certainly not lacking in tragic and humiliating memories. Although it is impossible to quantify how social pathologies contribute to the demographic imbalance, one cannot simply exclude them from the explanation, more so because the conflict in Northern Ireland involves Eire as well, not only for empathy with the struggle in the North, but also because the Republic's territory has been attacked several times by Unionist terrorists.

According to Coward, between 1985 and 2000, a 19% population growth is to be expected in Eire, against 8% in Northern Ireland, and 1% in Britain. In his paper "Demography: the 1980s in perspective," Compton attempts a projection of population trends by means of a comparison between Catholic baptism registrations published in the *Annuario Pontificio* and birth records of the Registrar General: the Catholic birthrate is thus estimated between 18 per thousand and 20 per thousand while that of Protestants falls between 14 per thousand and 15 per thousand. As in the whole United Kingdom, the birthrate in Northern Ireland fell to a minimum in 1977. After that year, a rapid growth was expected, as the plentiful offspring born in the Sixties got married and had children. Such a rise has not occurred, and this leads Compton to argue that during the Eighties a substantial fertility decline has taken place. In the Republic the birthrate plummeted from 21.9 per thousand to 14.7 per thousand between 1980 and 1989, and was therefore not very different, at that time, from that of the Northern Irish Protestants.

Compton rejects the hypothesis that conflict might in some way influence the Catholic birthrate, keeping it high and leading the Catholics to achieve majority status. He interprets the present trends as part of a diffusion process whereby the fertility of the Northern Irish Catholics should decline in the future, with a lag of about ten years after the Protestants, so that the Catholic population could reach equilibrium at a level slightly below 50% of the total. Protestants will therefore retain a majority, albeit a marginal one. It is a matter of time before this opinion is proved to be correct or mere wishful thinking.

Coherently with his Unionist stance, Compton, in a paper entitled "Employment differentials in Northern Ireland and job discrimination: a critique," stresses once more his denial of the existence of anti-Catholic discrimination. According to him, this has disappeared, as there is no longer a local Protestant power, so that the present Catholic disadvantage in terms of career and wages is due merely to labour market mechanisms. But the supporting evidence is ambiguous, and in many cases can be easily used to demonstrate the opposite.

Again, the historical perspective is wanting: labour market mechanisms have been severely distorted. If Catholics live in peripheral areas or in ghettoes where they have hardly any chance of finding a job, who has compelled them to become segregated in underprivileged neighbourhoods? If they prefer not to look for a job elsewhere in Northern Ireland, who "discouraged" them? And how, by what threats, or locked doors and threatening graffiti near the entrance of firms, have they been "persuaded" that it was better not to risk asking questions about vacancies? Paul Hill, one of the Guidford Four, tells in his autobiography *Stolen years* how he was discouraged to apply for a job precisely by Unionist graffiti.

If Catholics are proportionally more numerous in the lower classes whose employment prospects are bleak, who has pushed them down the social ladder, not yesterday, but for centuries? Catholics find jobs in the imperial services such as the Royal Mail, but are still largely under-represented in local services, such as the municipal administrations, quite often still under Unionist control. And it is only since 1972, under the direct British administration, that some efforts to remedy past ravages have been undertaken.

That said, however, it is fair to acknowledge that under the new administration, the social structure in Northern Ireland is changing. A Catholic middle class is rapidly growing and rising, thanks to educational advances and a housing market which is gradually losing, for those who can afford it, its discriminatory character. New, less rigid cleavages are therefore emerging, based upon social stratification rather than upon Church affiliation.

About 40% of Northern Irish university students attend British Universities, and more than two thirds of them are Protestant. About half of the Catholics come back to Northern Ireland after finishing their studies, while only one third of the Protestants return. This Protestant brain-drain opens up for Catholics better study opportunties in the two universities of Northern Ireland. There is a decided trend inversion here, in comparison with previous patterns, when comparatively more Catholics than Protestants emigrated. The student body at the Queen's University in Belfast was 70% Protestant in the late Sixties, while at present Protestants are merely a half of the total.

A Catholic middle class had been in existence for some time. It was mainly made up of medical doctors, lawyers, builders, and shop and pub owners, but was a comparatively restricted group. At present this group has grown briskly, has become diversified, and has invaded professional fields which had long been the preserve of Protestants. This process had already begun in the years immediately following the Second World War, when a number of reforms by the government of the United Kingdom (certainly not by the Stormont regime) put higher education within the reach of more and more people.

The new Catholic middle class usually votes for the SDLP, and has become gradually detached, even psychologically, from the more radical lower classes who keep supporting Sinn Féin and look at the *nouveau riches* not without envy and antipathy. Members of this Catholic middle class find a direct British administration advantageous, tend to create a niche for themselves in the present system, and are hostile to violence, not only from Unionist paramilitaries, but also from the IRA.

Malone Road and the surroundings — the most expensive zone of Belfast — in the past a Unionist stronghold, is swiftly filling up with Catholics who already make up half, perhaps more, of the total residents. With acrid sarcasm Protestants call the area "Vatican City." The local Catholic church of St. Brigid, built towards the end of the nineteenth century for the servants of the great houses, has now more than 5000 parishioners, most of whom are owners of the house they live in.

Street signs in wards inhabited by the new Catholic middle class are in English, whereas in Catholic low class areas, bilingual English/Gaelic signs have appeared. It would be a serious mistake, however, to regard such an emerging middle class as less nationalistic. Satisfied that his social, religious and ethnic identity is no longer so harshly discriminated as in the past, the well-to-do Catholic avoids ethnic ostentations, but remains aloof from the Unionist camp.

CHAPTER 6

REGIONAL PLANNING AND HOUSING IN AN EMBATTLED SPACE ECONOMY

Partition is not without economic consequences. What is missing in the Irish space economy as a whole is a strong development pole. This is largely due to a lack of hardly replaceable economic potential, and more precisely, to the smallness of the internal markets in both Irish statelets. The frontier between the two Irelands is not an impassable barrier: after all they are both within the European Community. But Northern Ireland, in its present condition, is a burden to Great Britain and definitely not a source of support to Eire.

Northern Ireland as a whole is a problem region, with a very high unemployment rate since the First World War. The rate has stayed constantly higher than that of any other region of the United Kingdom since 1936. Catholics are by far overrepresented among the unemployed and in the humblest jobs.

Regional planning in Northern Ireland was started rather hesitatingly with the partial application of the Planning and Housing Act of 1931. Some early attempts of the regional government to attract industries in order to tackle unemployment took place in the Thirties with the New Industries Acts of 1932 and 1937.

During the Second World War the debate on planning intensified. However, determined local opposition by politicians and entrepreneurs soon caused the wholesale stagnation of planning activity and delayed for many years the introduction of the innovations proposed, among which was the setting up of a central planning agency. The region never had anything comparable to the Town and Country Planning Act of 1947, which played such a decisive role in land use control in Britain. In Stormont, constantly dominated by the Unionist Party, there never was any political support to planning such as that given in Westminster by the Labour Party.

Consequently, the region "skipped" the whole development phase of British planning in the Fifties.

Between 1960 and 1972, British experts, briefed by the regional government, drafted some plans, which were short-lived and disappointing. After the introduction of direct Westminster administration in 1972, the enactment of a Planning Order brought forth a concentration of powers unique in the history of planning in the whole United Kingdom. In fact, since October 1973, all powers on this matter were taken up by the Secretary of State for Northern Ireland.

On the strength of the New Towns Act for Northern Ireland of 1965, enacted nineteen years after its British namesake, the construction of the new town of Craigavon, Co. Armagh, was decided upon. After that, some regional centres were designated: the towns of Antrim (1966), Ballymena (1967), and Derry (1969).

The Belfast Regional Survey and Plan of 1963, aimed above all at the development of urban infrastructure, including urban motorways, was hotly opposed, especially in the Protestant-Unionist stronghold of Shankill, where the project was regarded as a trick to break up local neighbourhoods. Reactions of this kind, in a context such as that of the Belfast ghettoes, underpin an eminently static society. The neighbourhood is seen as a protective shell against an external "hostile" world: the "threat" comes from the neighbouring Catholic ghetto of Falls, but also from a government authority regarded as "too remote." Any change in the lazy circle of habits and ideas is regarded as a menace. As pointed out in Chapter 4 it is in such conditions of social and psychological stagnation that anti-Catholic pogroms are hatched.

The plans for the motorway, which would have brought about the razing of a part of Shankill, were not implemented in the face of intimidation from Protestant paramilitaries. In 1982, under direct British rule, a new decennial plan for Belfast was formulated. In agreement with Tory government policy, this plan leaves greater room to free enterprise.

Little effort was made to assess whether the British tradition was appropriate to the conditions in Northern Ireland nor the whether the British planning itself had yielded the expected results. Northern Irish planning practice is, in fact, still focussed upon measures aimed at the control and decentralisation of development adopted in Britain in the buoyant conditions of the Sixties, and was not particularly successful then either.

Accordingly, the Northern Ireland Regional Physical Development Strategy for the twenty year period between 1975-95, was aimed at decentralisation from Belfast and a policy of peripheral "growth centres:"

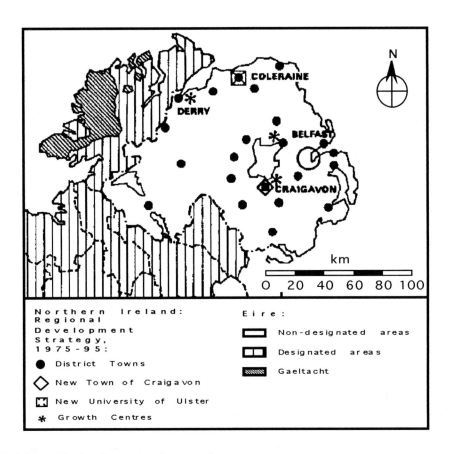

Fig. 6.1 — Regional planning framework.

Craigavon, Antrim and Derry (Fig. 6.1). As is often the case with territorial plans, while proposed remedies and results are not particularly satisfactory, the assessment of problems is unexceptionable. The disadvantage in terms of population size is correctly stressed. In fact, with only 1,577,836 inhabitants in 1991, Northern Ireland has, in demographic and market terms, a far more limited size than any other plannig region of the United Kingdom. Moreover the region is isolated from the large economic and population concentrations of Britain and Western Europe, while it possesses some of the features of the most depressed areas in Europe: eccessive presence of obsolete industrial sectors whose manpower requirements are declining, extreme scarcity of jobs and consequent unemployment and emigration, low incomes, and a highly dispersed rural population.

Unlike the other problem regions of Western Europe, Northern Ireland is embroiled in a conflict which surely does not provide encouraging conditions for potential investors. Only the Basque Provinces of Spain are experiencing a similar condition of ethnic tension and civil strife.

Awareness of the dearth of jobs, as well as the insecurity brought about by the conflict, discourage the unemployed from adopting the most obvious measure of moving around in Northern Ireland in search of a job. Usually the search for jobs is carried out locally, the only alternative solutions being either to live out of unemployment subsidies or to emigrate overseas.

It is noteworthy that even the 1975-95 Strategy mentioned above, though formulated by the Department of the Environment under direct British rule and no longer under the Unionist regime, does not offer particular incentives in the west of the region, which is the most disadvantaged part, and has the strongest Catholic presence. No attempt is, therefore, being made to tackle the problem of the regional imbalance within Northern Ireland, which has been worsened in the past also by discriminatory policies of the Stormont regime, whose consequences are still felt today.

Such are, for example, the cases of the new town of Craigavon and of the second University at Coleraine (see Fig. 6.1), both sited in such a way as to studiously avoid providing any support to areas with higher Catholic concentrations.

The proposal for Craigavon is contained in the cited Belfast Regional Survey and Plan of 1963: the existent towns of Lurgan and Portadown were expected to unite and expand forming a substantial new town of about 100,000 inhabitants. All other urban centres for which intensive development was envisaged (Antrim, Ballymena, Bangor, Carrickfergus, Downpatrick, Larne, Newtownards), named for the occasion "expanded towns," were, and still are, under firm Unionist control. Derry, then the second town of Northern Ireland (recently overtaken by Newtownabbey, belonging, however, to the Belfast urban area), with its mainly Catholic population and a Unionist control achieved only through shameless gerrymandering (until 1973), was instead conveniently ignored, in order to avoid providing jobs for Catholics.

For the same reason, Coleraine and not Derry was chosen in 1965 for siting the University of Ulster, the second in the region after the Queen's University of Belfast (founded in 1845 as one of the Queen's Colleges and raised to University status in 1908). The Unionist regional M.P., Nixon, of the North Down constituency, stated that a member of the Unionist government had informed Nixon himself that nameless, faceless men from Londonderry had gone to Stormont and advised against the siting of Ulster's second University in the city or settling industrial development there.

Nixon's statement was supported by Alderman Austin of Derry (*Belfast Telegraph*, 7th May 1965). The members of the pressure group were also named by the newspaper: they were prominent local members of the Unionist party or of the Apprentice Boys of Derry (a Protestant Association set up to celebrate an episode of the war between James II and William III of Orange).

The president of the Chamber of Commerce of Derry called for a petition demanding an inquest on the allegations, but he withdrew his proposal the following day, disclosing that he had received death threats. Nixon was sacked by the Unionists, and the party organization of his constituency passed a motion against him. The inquest was never carried out, the matter was never properly cleared up. But it stands to reason that when people wishing to have the matter investigated receive death threats, the worst suspicions are abundantly confirmed.

After the introduction of British direct rule, the Department of the Environment sought to tackle the overcrowding in Belfast ghettoes by building new housing estates. These attempts were met, however, with harsh opposition by the Protestant-Unionists, fearful of a territorial expansion of Catholics which might upset local political majorities.

Further examples of economic and territorial discrimination concern ports: the only comparatively modern and efficient ones are those of Belfast, whose port concentrates between half and two thirds of the Northern Irish cargo; and Larne, which handles most of the rest. Needless to say, both Belfast and Larne are in areas with strong Protestant majorities. The ports in predominantly Catholic areas (Derry, Newry, Warrenpoint) have received but scanty investments, and as a consequence are very poorly equipped: collectively, they handle less than a tenth of the aggregate cargo.

Northern Ireland is regarded as a particularly disadvantaged region by the European Community. Between 1975 and 1981, it received from the Regional Development Fund, over 112 million pounds for industrial and infrastructural projects, and about 123 million pounds between 1973 and 1981 from the European Social Fund for the promotion of training and manpower mobility, and the creation of new jobs. Further support has been forthcoming from the Agricultural Fund and by the European Investment Bank. Even before the United Kingdom became an EEC member, the region already had extensive social welfare and training programmes in line with the British Welfare State tradition.

In April 1982, following the example of the IDA of Eire, the Industrial Development Board for Northern Ireland (IDB) was set up. The IDA, however, notwithstanding its rather poor performance, seems to operate in more favourable conditions. In fact the Republic, unlike Northern Ireland, was not encumbered by a large number of obsolete manufacturing establishments, and could therefore start from greenfield site conditions, besides possessing the advantage of political stability, except for the occasional terror attacks from Northern Irish Protestant loyalists.

Inducements offered by the IDB were very high: grants for covering capital expenses (buildings, machinery and equipment) could be up to 50% of the total investment, while in the more strongly assisted areas in Britain, the Special Development Areas, financial inducements never went over 22%. In 1988, the Standard Capital Grant was replaced by a programme of Selective Financial Assistance.

Industrial jobs plummeted in Northern Ireland from 177,000 in 1970 to 95,000 in 1982, then rose slightly to 101,000 in 1984. In 1993, industrial jobs were down to 93,000. The IDB has been compelled to admit that a large number of industrial firms need assistance merely to survive. Nevertheless, government prompting of manufacturing development may have at least succeded in dampening the impact of the recession, helping enterprises to keep their investment levels (Sheenan).

The disadvantage of a peripheral location seems, therefore, to be slightly decreasing. This, however, is probably the result not of government planning, but rather of more general spontaneous trends, whereby, since the 1981 Census, and even more since the 1991 Census, the more peripheral and remote localities in the United Kingdom, such as those of the Scottish Highlands, have proved to be the most dynamic, at least in demographic terms. This may be a consequence of the drive towards "unpolluted" environments, "far from the madding crowd," coupled with the steadily improving circulation of information, typical of the post-industrial society.

Though the United Kingdom is approaching a post-industrial stage, Northern Ireland is certainly far from such a level of development. Its employment structure in 1991 might indeed suggest the picture of a post-industrial society, with only 3.5% of the employed in primary activities (agriculture, forestry, and fishing), 22.2% in secondary activities (construction and manufacturing), and 74.3% in services. This quite misleading impression is quickly dispelled, considering that 43.2% of the employed are in public administration and "other services," thus providing a picture in which *deindustrialization* and *assisted economy*, rather than "post-industrial economy" are the basic keywords. A third basic keyword is, of course, *unemployment*. Official unemployed claimants are only 13.9%, but the actual figure is probably higher, as the employed in 1993 were only 544,000 (274,000 males and 270,000 females), that is to say little less than half the economically active of working age (16 and over), numbering 1,184,000.

As in the Republic, the response of a great many people to the bleak economic situation has been emigration.

The economic predicament of Northern Ireland as a whole is compounded, of course, with that of the western rural areas. As in the Republic, these areas are very marginal, and suffer severe economic and social problems, further sharpened by territorial policies.

This point is illustrated, for example, by the wholesale closure of elementary schools, especially in the Sixties. Even shools with three teachers have been closed. This process of rationalization was economically justifiable, at a time when radical changes were in the offing in the school syllabus, with the introduction of new subjects such as environmental studies, art and physical education, all of which require highly trained teachers who are not always eager to take up posts in particularly isolated peripheral areas. Nevertheless, all this means considerable drawbacks for an already disadvantaged population. The provision of free bus transport to pupils living more than 2 miles (3.2 km) away from the nearest school has only proved to be a very partial alleviation.

Even worse negative effects have been wrought by the centralization of medical services, with the attendant closure of many rural dispensaries. The demand for these kinds of services is not so massive and predictable as that for school services, while at the same time it can be far more urgent and a life-and-death matter.

Not only the public sector, but also the private one has contributed to worsening rural accessibility. From the Sixties onwards, a rush towards the rationalization of retail outlets has taken place, with a spate of closures of small country shops, and the building of massive centralized shopping centres on the periphery of urban areas. The comfort of the comparatively affluent and mobile has thus increased, while households without cars, whether due to poverty or old age, are experiencing serious deprivation.

The public transport system (Ulsterbus) provides a substantial coverage of the territory, but accessibility levels remain highly variable, with several small scattered areas of deprivation in the west, and even in the east, especially in Co. Antrim, Co. Armagh, and in the south of Co. Down (Jordan & Nutley).

Are the present development trends sustainable? The fashionable concept of *sustainable development* is not easy to define as it hinges upon socially constructed values, with their inevitably subjective underpinnings and ambiguity. Yet, in spite of that, or perhaps just because of that, the concept is vital and stimulating, while more precisely defined theoretical constructions are weaker just because they seek to attain an elusively high level of precision. The concept is commonly taken to indicate environmental sustainability, but nothing prevents us from considering economic and social dimensions as well.

From a strictly *environmental* viewpoint, Northern Ireland is not confronted with very serious problems of sustainability. Its comparatively weak population density (in Western European terms) of slightly over 100 inhabitants per square kilometre, the virtual absence of mining activites, the definite trend towards the demise of heavy industry, the abundance of fresh water, the frequent winds sweeping away air pollution, all contribute to a comparatively clean environment. The only problems seem to be due to localized areas of pollution in the heavily populated Belfast urban area, some vulnerability of water resources, the need to protect some endangered species, and the character of some particularly valuable environments which could be threatened by unregulated tourism, as well as by the alteration of riverine environments discussed in Chapter 1.

But what of *economic* sustainability? Here the problems appear to be far more serious, as without the support of the United Kingdom in terms of economic incentives and welfare structures, Northern Ireland would sink to a far lower standard of living.

But it is under the viewpoint of *social* sustainability that the prospects are the bleakest, for obvious reasons, while the festering conflict is allowed to continue unresolved. In no other field of planning is this so evident as in housing provision.

The most common territorial strategy, in any ethnic conflict, is that of segregation, freely sought or forced, to ward off real or imaginary perils. It may be practised by one of the rival groups or by both. People tend to stay with their own group, and sometimes intimidate "the others" in order to expel them from "their" neighbourhood.

This is precisely what happens in Northern Ireland, especially in the larger centres, where segregation patterns exist at least since 1911, and almost certainly even before. Larger Catholic communities tend to be more segregated. Segregation is stricter in the more populous towns. Smaller centres, besides having lower segregation levels, are less frequently affected by terrorism and have a higher percentage of the rare, but increasing, mixed marriages. However, as stressed before (Chapter 4), the tension is not usually relaxed by such marriages.

Even where segregation is less evident, there are innumerable subtle forms of discrimination and a tendency to reciprocal avoidance. Protestants go to certain pubs, Catholics to others. Gangs of Catholic youth regard some streets as "their own," Protestant gangs "control" other territories, and there are no mixed gangs. Meeting opportunities between members of the two groups are mostly in the workplace, especially where the discrimination in job allocation has not excessively distorted the labour market and the encounters can thus take place on a comparatively even footing. But this is a rare occurrence: the large firms, such as, for instance, the Belfast shipyards, are, almost without exception, a Protestant reserved domain; and discrimination is difficult to eradicate in Protestant dominated local authorities (see Chapter 4), even though it has been prohibited in the Northern Ireland Constitution Act of 1973, under direct British rule.

A more effective influence on phasing out discrimination has been achieved, from the Eighties onwards, in American private companies with subsidiaries in Northern Ireland, thanks to the implementation of the Séan MacBride principles which are aimed at American-style affirmative actions to provide employment to Catholics. These policy guidelines are closely akin to the Sullivan principles on the hiring of blacks in South Africa in the apartheid age.

While the Protestants use their power to discriminate — unchallenged power until 1972, areas of local power after that year — the Catholics have adopted, in self-defence, an avoidance strategy, through self-sufficiency. Charity and social work are centred around the Catholic Church through societies such as the Saint Vincent, and the educational system is likewise under Church control. Further instruments of Catholic self-reliance are the building cooperatives and autonomous industrial development schemes such as the Tyrone Crystal Factory in Dungannon.

The rigidity of Protestant closure to any possible cooperation with Catholics amply justifies this attitude. It must be borne in mind that, in half a century of unchallenged Unionist regime from 1921 to 1972, only one bill of the opposition has ever become law in the Stormont parliament: the Wild Bird Bill of 1931 for the protection of wildlife; evidently a problem regarded as utterly irrelevant for the balance of power.

One of the most controversial features of Unionist territorial policy is, as we have seen, the housing policy. In June 1968, Austin Currie, an opposition

member of the Stormont Parliament, "illegally" occupied an empty council
house in the village of Caledon, Co. Tyrone, in order to attract attention to
Unionist discriminatory practices. A short time earlier, the allocation of a
council house to a Protestant spinster, while many large Catholic families
were left stranded, had caused a sensation. Currie's action was intended
precisely as a response to abuses of this kind. The incident led to the first
civil rights marches and thereby, as we have seen, triggered Protestant
persecution against Mr. Currie and his wife, and heralded the beginning of
"the Troubles."

The government of the United Kingdom identified the misuse of council
housing allocation for purposes of political-territorial control as one of the
major friction points in the social life of the region (Cameron Commission),
and sought in earnest to redress the imbalance. Between October 1971 and
July 1973, a new Northern Ireland Housing Executive (NIHE) took over the
direct control of council housing.

The NIHE is an authority entrusted with decisional powers, whose
executive board is made up of nine members: three nominated by the
Northern Ireland Housing Council and six, including the chairman,
nominated by the United Kingdom minister responsible for housing. The
NIHE therefore became the largest regional housing authority in the British
Isles. Initially, it effectively succeeded in promoting housing construction.
In 1971, the number of flats built per 1000 inhabitants was 9, but the figure
dropped rapidly: in 1980, it was halved by the unrest and the activities of
opposing paramilitary groups. Between 1981 and 1987, over 64,000 houses
were completed, or about 9,200 per year (32.4% by the local authorities,
62.8% by private builders, 5.8% by associations and "others"), or 5.9 houses
per 1000 inhabitants. Especially in Belfast it was increasingly difficult to
persuade building contractors and workers to operate in the "hottest" areas,
precisely where the need is the greatest.

Further problems are brought about by confessional divisions historically
entrenched in the building sector due to the traditional discriminatory
strategies and the reciprocal avoidance of the two communities. For
example, in a given town, the masons may be all Catholic and tilers all
Protestant, so it is hard to prevail upon such a divided manpower to work
together: in times of unrest men tend to refuse working outside the areas of
their own group. Moreover, it is exceedingly hard to find building plots, due
to Protestant opposition to any expansion of Catholic territory.

Complaints of discrimination in housing are at present voiced rather by
Protestants, who object to an "excessive" share of available resources being
invested in Catholic areas. They neglect the fact that this unequal allocation

is due to the massive backlog accumulated in the housing of Catholics who for such a long time were discriminated against in resource allocation, and so often violently ejected from their dwellings by Protestant extremists.

Yet, notwithstanding the theoretically impartial criteria in housing allocation, the NIHE has been obliged to accept, especially in the major centres, a rigid zoning akin to that of the old South African Group Areas Act. Often the allocation of housing has been decided by the paramilitary groups and, given the imbalance of strength between the two communities, the main effect, in most of Northern Ireland, has been to keep Catholics within their increasingly insufficient ghettoes.

The difficulty in finding new building plots prompted the enactment, in 1976, of the Housing (N.I.) Order, which introduced the concept of the Housing Action Area with the emphasis on relieving housing and social stress through the rehabilitation of existing buildings rather than redevelopment. A Renovation Grants Scheme aimed at improving private housing conditions by encouraging home-owners to refurbish their homes was also introduced.

Nevertheless, social and housing conditions, both for the long neglected Catholic communities, and for the less affluent Protestant classes, are among the worst in the British Isles. The belated NIHE efforts have not yet succeeded in bringing about decisive improvements.

In 1974, there were 90,000 "unsuitable" houses due to insanitary conditions and precarious services (without runnig water and/or sewage, or in disrepair), that is to say 19% of the overall housing stock in Northern Ireland, and therefore, four times the percentage in England in 1976. Obviously, the percentage of officially unfit dwellings was far higher in the predominantly Catholic areas of southern and western Northern Ireland, as well as in the Belfast Catholic ghettoes.

Between 1974 and 1979, unfit dwellings dropped by about 25%, but this decrease was mainly due to demolitions, while other dwellings have, in the meantime, become unfit in greater number than those which were refurbished, with an yearly average of 4,600 against 3,800. In 1979, the percentage of unfit dwellings was still three times higher than in England, and there were 32,000 families on waiting lists for a new house, while 14,000 dwellings were destroyed or damaged by bombs in Belfast alone. The building of new housing has also been slowed down by the allocation of time, resources and manpower to rebuilding, or repairing, damaged dwellings. Between 1981 and 1992, the rate of completion of new dwellings has dropped from 2 to 0.7 per 1,000 population. Significantly, while dwelling prices have decreased in England by 10% from 1990 to 1993, they have grown in the same period in Northern Ireland by 14%.

With the exception of a few middle-class suburbs, such as the Malone Road area in Belfast discussed in the previous chapter, residential segregation, already very high, has grown in traumatic fashion since the onset of the civil war, in all urban areas, particularly in Belfast, Derry and Craigavon. In the latter town, a directly negotiated exchange of dwellings between Catholics and Protestants has taken place. In Derry, where Catholics are decidedly the majority, a flight of Protestants from the historical centre sited on the west side of the river Foyle, has taken place towards the eastern suburbs where the Catholic population is sparse. In Belfast, on the other hand, where Protestants are in the majority, delocation has affected above all Catholic families: between 1969 and 1973, more than 10,000 dwellings were relinquished, and about 60,000 people, mostly Catholic, obliged to seek shelter in the ghettoes of their own group.

Terror attacks, and above all the car bombs, have seriously affected commercial centres in the towns, as we shall discuss in greater detail in the following chapter. Obviously, the civil war has worsened unemployment, which has soared above 20%. A few firms have dared to invest in Northern Ireland, attracted by the windfall of governmental incentives. These firms, in general, have chosen sites well inside "secure" areas, that is to say, those with a smaller Catholic population. This has contributed to worsening the social and spatial imbalance.

West Belfast, for instance, where most of the Catholic population of the city is concentrated, is entirely devoid of industries and affected by overcrowding due to the inflow of refugees. It has little capital to promote self-reliant initiatives, unlike Catholic areas of other urban centers in less depressed socio-economic conditions. The unemployed from West Belfast suffer from serious mobility problems in their search for jobs as their section of the city is surrounded by hostile Protestant areas.

Summing up, then, besides the obvious physical destruction and human casualties and suffering, the territorial consequences of the civil war are probably the following: (i) growing segregation, (ii) worsening housing conditions in spite of the efforts of the impartial housing policy belatedly introduced by the British, (iii) objective difficulties encountered by the same housing policy, (iv) the downgrading of retailing in urban central areas, (v) increasing unemployment, and (vi) worsening social and economic imbalances.

CHAPTER 7

A DIVIDED CITY

The very birth of Belfast, a "gateway" for an alien colonising population into hostile territory, is, by definition, bound to a conflict process. The same applies to all other cities in Ulster and in Ireland in general.

Belfast is located in a typical privileged site, being the easiest gateway into Ulster and one of the best in the whole of Ireland. It is well sheltered in the terminal section of the Belfast Lough, the drowned part of the Lagan valley, and enjoys easy access to central Ulster through the natural corridor of the Lagan, about 7 kilometres in breadth, set between the Eocenic basalt cliffs of County Antrim to the north and the Silurian hills of County Down to the south. The role of this corridor in the development of the region cannot be overemphasized.

In 1613, the settlement was granted town status. Around 1640, the city began to overtake Derry and Carrickfergus in shipping activity. A map of 1685, drafted by Thomas Phillips, the royal cartographer, shows the town walls and the rows of cottages along the roads outside the walls in a westward direction. These were the dwellings of the Irishmen, while the houses inside the walls were reserved for the settlers from Britain: the typical set up of a colonial town under threat of the dreaded native uprising.

By the dwellings of the Irishmen, outside the walls, there were two mills close to the Farset brook, a left hand tributary of the Lagan. This area, at present at the confluence of Mill Street and Divis Street, was to become the nearest point to the city centre of the Falls area, at the meeting of the roads coming from the rural spaces to the west of the town. To the north of the Falls area, the Protestant area of Shankill was taking shape at the confluence of roads from the Antrim countryside, whose population was, and is, Protestant. The area of the two mills became the centre of the future industrial zone of the town. Therefore, already in the late 17th century, the foundations were laid for what was to become West Belfast: a place of industrial suburbs and ghettoes, rife with potential conflict.

During the 18th century, a key role in the development of the town was also played by the port, which traded mainly with Great Britain, the West Indies and North America. The development of Belfast as a gateway was dependent on one side upon the *hinterland*, supplying raw material for the linen manufacturing and the food industries (butter, bacon, whiskey), and, on the other side, upon the *foreland*, that is to say the American colonial spaces, whence came mainly tobacco and sugar. The town's textile and food products were distributed throughout the hinterland or exported overseas.

Deforestation and population growth in the hinterland brought about road improvements, the replacement of the Irish cart with its massive wheels by the more efficient Scotch cart with spiked wheels (towards the end of the 18th century), the digging of the Lagan canal (1754-63) which eased communications with the rich Lough Neagh plain, the first infills to improve harbour accessibility (from the second half of the 18th century).

Flax cultivation and the attendant cottage spinning and weaving, spread largely in the hinterland. In the second half of the 18th century, Belfast became a significant centre of the brown linen trade. The first Marquess of Donegall, therefore, had the Brown Linen Hall built in 1773. Many other Ulster towns, however, such as Armagh, Ballymena, Cootehill, Derry, Dungannon, Lurgan and Newry, were also important commercial centres for brown linen. The greater part of sales occurred on a strictly local scale.

The beginnings of urban-industrial polarisation are due to an important innovation: the introduction of bleaching. In the 18th century, this process was carried out by immersion into alcaline solutions. It was no local invention as the technique had been in existence for some time in Western Europe. In Ireland itself, Dublin already had some production and marketing of bleached linen. However, it was a novelty for Ulster. Bleaching factories needed sizeable economies of scale to operate effectively, and they tended to concentrate in the immediate vicinity of Belfast. This polarising effect influenced the commercial sector too and, in 1783, the White Linen Hall was built in Belfast.

The quickening tempo of urban growth at the turn of the 18th century is linked also to the growth of the cotton industry, set up in Belfast in 1777 by the Charitable Institute to provide employment for the poor. Water power was used for spinning since 1784. The rise of this industrial sector increased manpower demand, so that in 1791, the youngsters of the Institute began to be hired as apprentices by the local entrepreneurs. The number of spindles in activity rose from 8,000 in 1790 to 50,000 in 1811.

In the first decade of the 19th century, the introduction of the steam engine in cotton spinning was a further innovation which promoted the concentration of economic enterprise and population in Belfast. In 1811, in the immediate surroundings of the town, fifteen steam engines were in activity. The mechanization of the cotton industry gave rise to powerful concentrations of production and manpower. The factory system was born, and the collapse of cottage manufacturing had begun, with very serious social consequences leading to the machine-breaking (luddite) movement. This, however, was felt especially in Nottinghamshire and other English counties, rather than in Ireland, where the growth of the cotton factory system was rapidly stunted.

After the mechanization of spinning, the factory system was just beginning to spread to weaving when, in 1825, a severe slump took place. While in Britain the robust industrial structure was far more resilient, the Irish cotton industry was unable to withstand competition.

The technological know-how achieved in the cotton industry, however, proved very useful in the mechanization of the linen industry, in which Belfast had no competition to fear. The first linen spinning factory was that of Thomas Mulholland, who replaced his cotton spinning mill, destroyed by fire in 1828, with a linen mill. In 1835, there were already ten linen spinning mills in Belfast, of which two were converted from cotton. In 1860, there were thirty concentrated in the western part of the town, of which thirteen were in the Shankill area, the largest working class Protestant quarter. Entrepreneurs had shunned the Catholic ghetto of Falls, a cheap manpower reservoir, but devoid of industries, not unlike the black ghettoes in old apartheid South Africa.

Of course, in South Africa, both the ghettoes and the avoidance of them by entreprenurs were prescribed by the laws of the country, whereas in Northern Ireland such things were a matter of custom. It seems an important difference. Yet the results were similar.

Industries linked with linen included foundries and engineering for textile machinery. There was also the development of a production of sulphuric acid for bleaching, as the swifter acid process began replacing the alcaline one from the first half of the 19th century. Belfast became a node of accelerated change in which cumulative and self-reinforcing processes tended to support its growth and strengthen its dominance on the surrounding area.

These processes, typical of a dynamic centre, were due to a number of feed-back effects (Friedmann): (i) the weakening of the peripheral economy through net transfer of resources from the periphery to the centre (*domination effect*); (ii) the rising circulation of information, thanks also to the growth of the population, production and income, whereby the probability of innovations increased (*information effect*); (iii) the creation of intellectually stimulating conditions favourable to innovations (*psychological effect*); (iv) the emergence of new social and economic goals and the transformation of behaviour patterns and institutions (*modernization effect*); (v) the tendency of innovations to give rise to further innovations (*linkage effect*); (vi) the effective exploitation of innovations by means of economies of scale (*production effect*).

This is a clear example of the close links between urban growth and industry acting reciprocally upon one another, both propelled by innovation and connected to physical conditions, as stated earlier in the discussion on the physical space (see Fig. 1.7).

Beside the leading textile sector, other industries existed in Belfast, for the manufacture of tobacco, flour, beer, whiskey, rope, soap, and starch. There were also some sawing mills. Like textile mills, these activities were concentrated in the western part of the town, close to the terminals of the railway lines: from Lisburn (opened in 1839) and soon extended to Lurgan (1841), Portadown (1842), and Armagh (1848); from Ballymena and Carrickfergus (1853); and from Dublin (1853). The stations of the three lines, built on the outskirts of the town, soon became development poles within the urban fabric: around these innovative structures, factories and working class housing were built, while the railways poured into the town more and more loads of immigrants expelled from the countryside by famine and lingering destitution.

The development of the port, however, faced severe difficulties. It was not an easy landing place. Like the whole North of Ireland, the Lagan valley and the surrounding hills had been massively eroded by ice during the different glacial events of the last 2.4 million years. This resulted in very gentle slopes and a winding river channel. When the lower reaches of the Lagan were drowned at the rise in sea level occasioned by the melting of the (so far) last great ice sheet, the Belfast Lough thus formed was inevitably shallow and with a winding submerged river channel. These physical conditions obviously called for intensive human intervention to improve the approaches to the port if a sizeable traffic was to be accomodated.

Before the Great Famine, the largest ships had to cast anchor more than two miles offshore and unload by lighter, a part of their cargo before approching the quays, a disadvantage which doubtless discouraged local shipping. The stagnation of the port was largely due to the behaviour of the Chichester family. The despotic dynasty actually opposed development, and not only because of mental laziness, but also because an "eccessive" urban growth would have threatened the control of the local M.P. seats by the family.

Arthur Young, who visited the town between 1776 and 1779, wrote:

> The town belongs entirely to his lordship. Rent of it £2,000 pounds a year. His
> estate extends from Drumbridge (.....) to Larne. 20 miles in a right line and is 10
> broad. His royalties are great, containing the whole of Lough Neagh which is I
> suppose the greatest of any subject in Europe (.....). The estate is supposed to be
> £31,000 a year, the greatest at present in Ireland.

The control by the Chichesters had solved none of the problems of the town, not even the basic one of water supply, which the Charitable Institute had taken charge of, from the 1790s, when the situation had already become critical. Only in 1840, when the power of the family over Belfast was on the wane, the Belfast Water Commissioners came into being. By progressing from pumping from wells to the construction of aqueducts, they eventually put an end to the sequel of typhus and cholera epidemics that had repeatedly plagued the town.

The power of the great family declined, as it became engulfed in ever growing debts, and also due to reforms (in the House of Commons in 1832, and in the Irish municipal councils in 1840) which started a democratization of the political system, breaking to some extent the grip of the great landowners: a process which was to be completed only in the second half of the 19th century. In Belfast, as elsewhere in Ireland, therefore, the old closed oligarchic corporation was replaced by an elected council.

Socio-political and institutional change eased economic enterprise. In 1837, Parliament gave assent to extensive improvements in the port structures. Between 1839 and 1841, the municipality carried out the cutting and dredging of a channel leading to the port. In 1847, a Harbour Commission replaced the municipality in the control of the works. In 1849, the Victoria Channel was opened.

The new works enabled shipping traffic to increase substantially, which in turn justified further technical improvements. In the first half of the 20th century, two new channels were opened. The infill of the bay went on apace. New land was reclaimed for productive uses and the building of new quays. Reclamation had already begun in mid-18th century, but only after the end of the Chichesters' domination could works start on a truly massive scale.

In 1822, the second Marquis of Donegall was steeped in debts whose total amounted to the huge sum of 217,000 pounds and was, therefore, obliged to undertake a wholesale reorganization of his estate. Till then, both in the countryside and in town, leases had been only up to twenty-one years, and quite often shorter. The system yielded high profits for the landlord, but discouraged investment by the tenants. The paralyzing effect of the seigneurial regime introduced by the conquest appears, therefore, quite evident not only in the countryside, but in town as well. The new conditions allowed longer leases or even indefinite ones; occasionally a tenant could even purchase his plot. In any case, greater security of tenure was an incentive to housing stock improvements.

At the death of the second Marquis, in 1844, the estate was still heavily encumbered. Immediately after, the Great Famine ravaged Ireland, bringing about insuperable difficulties even to estates that had been more efficiently managed. The Chichester estate had, therefore, to be entrusted, in 1850, to the Encumbered Estates Commission. During the following eight years, most of the plots, both rural and urban, were sold to tenants or speculators, thereby triggering the impressive growth of Belfast during the Victorian age. This caused a sevenfold increase in the population to about 350,000: an increase due, to some extent, also to the immigration of skilled workers from Scotland and Northern England. Housing was hurriedly built at a low standard, and the city took up a drab outlook. New rows of Georgian style houses rose later in the suburbs, especially after the Second World War.

A sizeable contribution to the economic growth of Belfast came also from shipbuilding. The first yard was set up in 1792 by Richtie, who later took his brother and another entrepreneur named McLaine as partners. In 1820, the firm Richtie & McLaine launched the first steam ship built entirely, including the engine, in Ireland.

The demise of the dominion of the Chichesters had a very positive impact also upon shipbuilding, as it allowed the improvements in port accessibility and port structures mentioned above, which in turn, made possible the building of iron ships: this being an excellent example of a linkage effect among innovations. The first iron vessel was built in 1838 by the foundry Coates and Young. The first yard for iron ships was set up in 1851. It was purchased two years later by Hickson, owner of the Belfast Iron Works, who, the following year, appointed E.J. Harland as a manager. The latter was pretty soon able to take over the firm, and operated with increasing success, taking Wolff as a partner. The Harland & Wolff yards covered in 1882, the entire Queen's Island, on reclaimed land: they are a typical example of external entrepreneurship, since Harland hailed from Glasgow and Wolff from Liverpool. Richtie too, was a foreigner: a Scotsman by birth. Harland & Wolff employed 500 workers in 1861, but peaked to 9,000 in 1900. Another shipbuilding firm, Workman & Clark, began business in 1880.

The shipbuilding sector linked up with foundries and engineering industries, thereby giving rise to vertical integration. This is an example of a typical industrial development process, which normally leads to the formation of clusters of technological and managerial innovations, and thus to cumulative growth, until market saturation is reached and a structural crisis becomes inevitable.

A structural crisis actually hit the whole United Kingdom in the late Victorian age, due to market saturation for some of the sectors (textile, iron and steel, and engineering) which had been the leading ones during the first Industrial Revolution, and because of foreign competition, as other countries followed Britain on the path to industrialization. The paleoindustrial economy of Belfast was severely affected. Within the spatial system of the British Isles, the city and all Eastern Ulster took on by degrees the features of a *declining transitional periphery*, suffering from unemployment and emigration. The onset of the Wall Street depression, from 1929 onwards, of course, did not improve matters.

The city, however, experienced a partial recovery thanks to investments from Britain in an industry which, at the time, had a good market potential. In 1937, an aeroplane factory was sited near the town, and two years later it provided 7,000 jobs. Industries in Belfast, however, became more and more dependent upon headquarters located elsewhere. A branch-plant economy was born. The poor local innovation potential and the narrow internal market of the six counties contributed to deepening the slump, although the growth of government services, when Belfast became the capital of Northern Ireland, provided some alleviation, at least for Protestants. Even as a capital, however, the city was still a tributary centre, and therefore, in any case, a periphery.

Unemployment was particularly severe among the lower classes. Catholics were hard hit. Textile workers decreased from 53,000 in 1926 to 43,000 in 1951. In 1934, the shipyards of Workman & Clark closed down, and only those of Harland & Wolff managed to survive the crisis. The private sector, no less than the public one, systematically discriminated against Catholics. While Catholic ghettoes became more and more overcrowded due to blatant discrimination in council housing allocation, the mainly middle class Protestant suburbs kept expanding eastwards, northwards, and southwards, while westward growth stagnated.

Demographic trends were similar to those of British industrial cities, with a rising pace of growth in the late 18th century, a spurt and rapid rise in the Victorian age, then a slow-down and inversion of the trend. In the case of Belfast, which plummeted since 1961 from over 440,000 to less than

300,000 inhabitants, there is, however, a considerable impact of the exodus towards suburban belt municipalities, largely caused by "the Troubles."

A social environment favourable to economic innovations, though not exactly on the forefront in generating them, Belfast, born as an instrument of Protestant dominion, ruled at all times by Protestants, has proved impervious to integrative social innovations in favour of Catholics. Wholesale bigotry and intolerance have been the typical Protestant attitudes towards the Catholic minority, so that the town has become one of the "hottest" areas of the Northern Irish conflict space.

The percentage of Catholics grew from 6.5% of the total in 1757 to 32.4% in 1834, and this increase escalated the fears of the Protestants, who unleashed the first anti-Catholic riots in the 1830s. In 1857 and 1864, the unrest took the form of true pogroms, and caused a increasing ghettoization of the minority, which in the meantime grew to 34.1% in 1861. The ensuing constant decline of the percentage to 23% in 1926, has multiple causes, discussed above, such as the impact of Protestant violence, the drying up of the rural source of Catholic immigrants after the exodus overseas, the inflow of Protestant immigrants from Scotland and England, the economic crisis which decreased the attractiveness to industries. Later, however, the Catholic percentage began to rise again. It is now around 40%, and therefore close to the percentage of Catholics in Northern Ireland as a whole. Such an increase of the Catholics is due not only to a higher birthrate, but also to the above mentioned heavy outflow of middle class Protestants to the suburban municipalities.

Segregation has been steadily increasing, especially after the pogroms which went hand in hand with the birth of Northern Ireland and, in 1969, with the onset of the unrest. In 1911, 41% of Catholics and 62% of Protestants were recorded as living in segregated streets. In 1972, these percentages had grown, respectively, to 70% and 78%. Segregation, in fact, is even higher than appears in the official data, both due to micro-segregation which escapes official surveys (a street seemingly integrated can be entirely inhabited by Catholics on one side and by Protestants on the other), and because the greater part of "mixed" areas are unstable and in transition, there being an inflow of Catholics and an outflow of Protestants, with the attendant expansion of Catholic ghettoes. The socially most prestigious areas, where owner occupancy prevails, are those in which segregation is comparatively lower, while the highest level is in the poorer areas, where council housing predominates.

A lingering conflict situation had not prevented, however, the emergence of a regular intra-urban hierarchy, with a decided dominance of the Central

Business District, largely coincident with the historical centre, and peripheral retailing and service centres of a lesser order along the main streets, which evidently acted as *space organizing arteries*. The conflict outburst in the late Sixties brought about a dramatic distortion of this intra-urban hierarchy, degrading the city centre and consequently increasing the importance of peripheral commercial poles. This was linked both to the flight of the middle and upper classes, and, therefore, of the purchasing power, towards the urban periphery or the neighbouring municipalities, and to the destruction of centrally located shops.

The Provisional bombing campaign gutted, between 1970 and 1975, about 300 central shops, approximately a quarter of the aggregate retail area. In the whole city, during those five years, about 1,800 explosions took place (in average one every day), of which 40% had retailing establishments as targets.

In 1972, the security forces set up rough barricades at the entrances of the commercial arteries of the city centre, later replaced by steel barriers with narrow entrances under strict surveillance. Parking of unattended cars was strictly prohibited in a much larger area around the walled centre to prevent car bomb attacks. Under improved local security conditions, contractions of the walled area have taken place in 1980 and in 1984. The barriers were restructured a number of times, and nowadays they even exhibit a certain decorative elegance, while the urban planning authorities of Belfast are expending efforts to create a positive image of the city (Neill).

Terror attacks arrested the process of rationalization and concentration of retailing, which in Belfast thus lagged far behind all the other major cities in the United Kingdom. A supermarket, in fact, is a far more substantial and "appetizing" target than a mere shop, and it is easier for a pretended shopper to plant a bomb in one. Moreover, restructuring of the retailing sector requires capital which seems hard to come by in such unsettled conditions. During the Eighties, however, propped also by suburbanization, the diffusion of supermarket and hypermarket chains has revived.

Before August 1969, the frontier between the Falls and Shankill ghettoes was, to some extent, open. After the onset of "the Troubles," it was almost entirely closed. Wall writing acquired a peculiar importance in territorial delimitation. Applying the language learnt from the old "western" movies, the territory of the other group is dubbed "Indian country" or "Comanche territory." Besides the graffiti, other visual aspects of the conflict are Orange Order marches, the drab outlook of ghettoes, razed buildings, the gaping gutted shops, and the security barriers (Photos 6-12).

Photo 6 — Loyalist symbols in South Belfast: a mark of territorial control.

Photo 7 — Lower Falls, West Belfast.

Photo 8 — A gutted building in South Belfast: the effect of an incendiary device planted in the shop on the ground floor by the IRA.

Photo 9 — A civil war landscape in Central Belfast: the effect of an IRA car bomb.

Photo 10 — One more building in Central Belfast razed by an IRA car bomb.

Photo 11 — Security barrier erected as a protection for the central business district of Belfast.

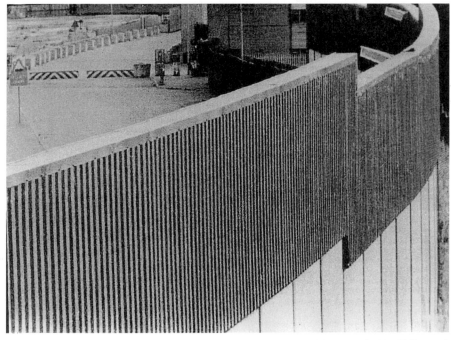

Photo 12 — The wall parting two ghettoes in Central Belfast: Catholic Falls and Protestant Shankill (From *De Telegraaf*, Amsterdam. Copyright Reuter).

As the authorities appeared unequal to the task of providing a modicum of protection, the communities of the various quarters organized their own defence, setting up patrols of "vigilantes," and their own public transport systems (the "black taxis") and information services by means of newsletters, bulletins, and local radio to support the morale of the "brave boys on the barricades."

The whole of Belfast stands in great need of social and physical restructuring. This has been carried out, to some extent, in two of the most depressed and conflict-ridden areas of Belfast, Falls and Shankill, though with an uphill struggle against local opposition. For example, the NIHE planned a centralized heating plant for both wards, which should have been sited in the Falls area. When the inhabitants of Shankill realized this, they vociferously demanded, and obtained, an independent heating plant. In Belfast, not unlike in other major centres, territorial competition is perceived as a vital necessity: planning documents are always minutely scrutinized to make sure that no territory is being handed to the "enemy."

By means of intimidation and violence, the paramilitary groups are able to paralyze any initiative. This surely does little to improve housing conditions, of which Belfast stands at the lowest levels among the major cities of the United Kingdom. In 1982, according to the NIHE, 23% of the families in Belfast were without exclusive access to a bathroom or a shower, and 24.2% without a sink. These percentages are more than twice those of Liverpool, which notoriously has the worst housing conditions among the major English cities. As in other Northern Irish centres, decades of discrimination in council housing allocation have caused a serious backlog in Catholic areas, and NIHE efforts are still far from solving the problem.

One of the most impressive efforts of this kind was the proposal for the new Poleglass housing estate, to be sited just ouside the Belfast district, in the neighbouring Lisburn district. The plan envisaged 4000 houses for 16,000-18,000 inhabitants. It was submitted to the Department of the Environment in 1973, and received assent the following year. No specific provision was made for the allocation of houses to families of a particular group but, given the huge Catholic housing backlog, the estate would inevitably be an almost exclusively Catholic area.

The Protestant majority in Lisburn soon staged a rabid opposition, which caused the Department of the Environment to delay the beginning of the works and, in 1976, to reshuffle the plan. The number of houses was halved. It appears that, underlying such a withdrawal, there were pressures brought to bear upon the Secretary of State for Northern Ireland by Unionist M.P.s in the Commons at a time when they held the balance between the Tories and the Labour Party.

The Protestants of Northern Ireland, however, had lost the nearly dictatorial power they held under the Stormont regime until 1972, and it is noteworthy that they were compelled to rely on struggle methods typical of a minority group previously used by Catholics at the time of the civil rights movement, that is to say marches, processions, public meetings, and letters to newspapers. When the Department of the Environment consulted the Lisburn municipality, the latter turned to obstruction tactics, until, in 1979, in spite of the fact that the Labour government had meanwhile been replaced by Mrs Thatcher's Tory government, the Department gave up the consultation procedure and granted the construction permit.

The building of the Poleglass estate started, though according to the halved project, in 1980. At the end of that year, by night, almost secretly, the first houses were entered. Already before the beginning of construction, Protestant families had begun to move from the neighbouring quarter of Dunmurry, triggering the transformation of Dunmurry into yet another ghetto, as land-hungry Catholics purchased or rented houses there. Not one but two Catholic ghettoes were thus created, according to the well known housing transition process tending to generate segregation and, therefore, feeding the cumulative social distress process made of: *discrimination — social pathologies — distrust of authorities — conflict — fear — discrimination.*

An excellent investigation of social distress was carried out, in the mid-Seventies, by a Project Team of the Queen's University in the Belfast Urban Area, including the 51 wards of the Belfast district, plus 50 additional wards belonging to neighbouring districts. Thirty-nine variables were considered covering such features as unemployment, subnormal children, overcrowding of dwellings, dependency, social class, dearth of school buildings, families without a car, shared dwellings, low educational level, school truancy, mortality caused by bronchitis, housing types, birthrate, Borstal male inmates, and so forth. The statistical technique known as factor analysis has pinpointed a number of wards suffering from a high incidence of social pathologies. The study cited, however, does not mention social pathologies as such but, more politely, "indicators of social need." All these wards belonged to the district of Belfast only, so that attention can be focussed on the 51 wards of this district, excluding the neighbouring ones.

Two distinct pathological syndromes were found by the research team: one was concentrated in central and western wards, with high unemployment, low income, state subsidies and large families; the distribution of the other syndrome partially overlapped the former one, and was typical of the central

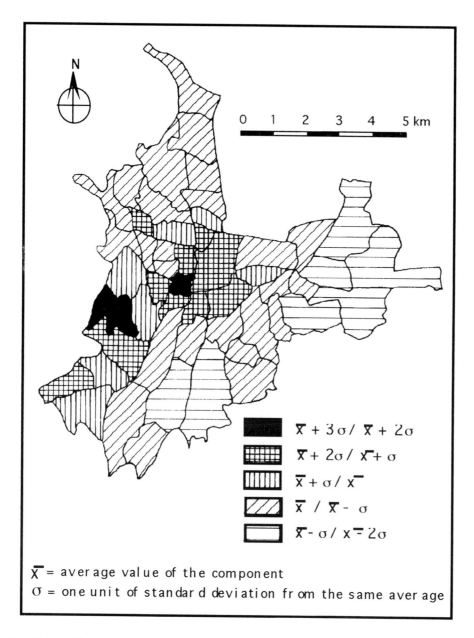

Fig. 7.1 — First component (accounting for 54.5% of the total variance) of principal components analysis of social pathologies in the urban area of Belfast. Explanation in the text.

wards alone — its main features were bad housing, environmental decay, low educational level, and personal handicaps linked to ageing or bad health.

Two central areas were selected by the team for an in depth-study: Lower Falls with the neighbouring area of Grosvenor ward, and the section of Shankill uphill from the area then being restructured; each with about 7,500 inhabitants in 1975.

Both areas were near to manufaturing establishments, then closed down or declining, which required scarcely trained and/or female manpower, they were served by many small corner shops and artisan workshops which could not survive physical restructuring and the competition of large scale industrial production. In the two areas, only one family in ten was not beset by at least one social pathology, while seven families in ten in the Falls and nearly five in the Shankill suffered more than one pathology; moreover 40% of family heads in the Falls had serious difficulties in understanding or completing bureaucratic forms of any kind, which obviously entailed severe social adjustment problems. It is clear that the Falls Catholics are in worse conditions than the Shankill Protestants, very likely as a consequence of discrimination.

In my previous book on Ireland, I reported the results of a simple principal components analysis, with twelve variables covering some social pathologies and the distribution of the Catholic population. The analysis was based on data from the 1981 census, concerning only the 51 wards of the Belfast district, which, as we have seen, include all those with a high incidence of socially pathological conditions in the whole urban area. The first component (Fig. 7.1), accounting for 54.5% of the variance reveals a latent dimension of juvenile dependency, housing distress, and a strong Catholic presence. This component is concentrated in the central and western wards, and thus shows their socially pathological conditions, thereby confirming the results of the Queen's University Project Team. The latter study, however, had depicted such pathologies without drawing any explicit distinction between the Catholic and Protestant areas, which is tantamount to glossing over the essence of the territorial conflict and therefore, over the cause of many of the present social and political difficulties.

CHAPTER 8

BEYOND NORTHERN IRELAND: DIFFERENT INTERPRETATIONS

Territorial conflict has, in Irish history, a constant and decisive role. As time went by, it took on the most different characters: it was rural or urban, it took place in a static society or while society was in a process of growing dynamic change, it raged on the battlefield or in the insidious form of guerilla warfare, with or without foreign intervention.

The conflict had two crucial turning points: the Reformation and the Great Famine. Reformation blocked the assimilation of the invaders. The Great Famine triggered a political movement away from Britain that, though by fits and starts, is not yet at an end. These statements are indeed at variance with a great many studies, recent and less recent. It is not surprising that issues so highly charged, emotionally and politically, should be subject to disagreements.

The role of the Reformation has been undervalued, but doubtless vast differences in the views of life divide Catholics and Protestants. This was, and is, especially true in Ireland, and particularly in Northern Ireland. Before the Reformation, Anglo-Normans and Englishmen tended to become *Hibernicis ipsis Hiberniores*, more Irish than the Irish, so much so that a kind of apartheid had to be introduced by means of the Statutes of Kilkenny of 1366, to prevent the assimilation of settlers by the natives. The Statutes were utterly ineffective as intermingling and celtisation continued unabated. Only after the Reformation did the settlers form a truly closed élite, the religious difference providing not only a ready means of identification but an effective ideological cleavage.

The key importance of the Great Famine has already been commented upon (Chapter 3) and does not need to be repeated.

Besides the analogies with South Africa, already mentioned on several occasions, there are important and very obvious differences. Ireland under the sway of the closed Protestant élite, after Cromwell's havoc and before the

Penal Era became less harsh — so roughly between 1650 and 1750 — was rather similar to apartheid South Africa. The emancipation of the Catholic middle classes under the leadership of Daniel O'Connell, the Young Ireland and the Fenian Brotherhood movements, the Great Famine, the giant migration overseas, and the uprisings which followed, are all typically Irish developments. The exodus of the young Bantu during and after the repression of the Soweto uprising from 1976 onwards, could, in a very limited sense, be compared, in terms of effects, to the Irish migration after the Famine, because both provided an external army, however in quite different historical circumstances and with different results.

The present conflict in Northern Ireland is also very different from the South African conflict: unlike the South African blacks until recently, the Northern Irish Catholics have been in possession of the franchise for a long time, and are still unable to fulfil their aspirations only because they are still a minority. South Africa is independent and, until a short time ago, politically isolated; Northern Ireland depends, politically and economically, on Britain whose government, in the present conditions, has declared that it is prepared to give it up to Eire as soon as the majority in the North is in favour of a united Ireland.

Structurally, however, the case of Ireland is akin enough to that of South Africa to justify the adoption of a system model of the conflict process which is very similar to that previously used for the South African case (Figs. 8.1, 8.2). In addition, the dependence of Northern Ireland on Britain requires an additional model of controlled conflict under an external superordinate authority (Fig. 8.3).

Within the traditional system, socio-cultural models and environmental factors favour population growth, which contribute to triggering the imbalance multiplier. Power and concentration of wealth, as a result of the foreign conquest also accrue to the imbalance multiplier.

The onset of the Industrial Revolution brought about innovations having many economic, social, spatial, and medical consequences, whose effects were felt especially in eastern Ulster and only in a subdued and indirect fashion in the remainder of the island. The imbalance multiplier concept is sketched in Fig. 8.1, as are the options open to the élite.

It must be borne in mind, however, that this élite is at least twofold: there is a British central élite and an irreconcilable peripheral élite made up of Northern Irish Protestant extremists. A third élite, that of Southern Irish Protestants was important during the golden age of the Ascendancy, but declined historically with the rise of the Catholic counter-élite and the approaching devolution of Ireland from the United Kingdom.

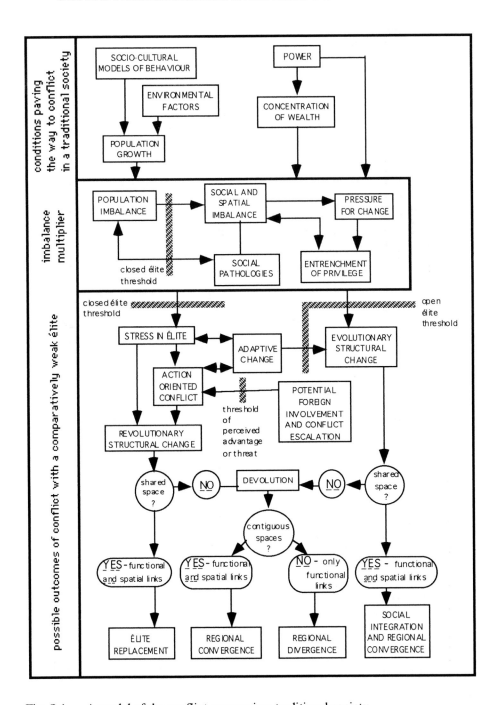

Fig. 8.1 — A model of the conflict process in a traditional society.

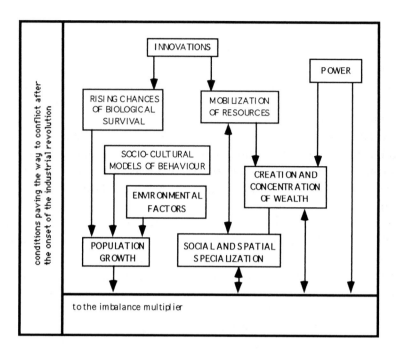

Fig. 8.2 — Conditions paving the way to conflict after the onset of the Industrial Revolution.

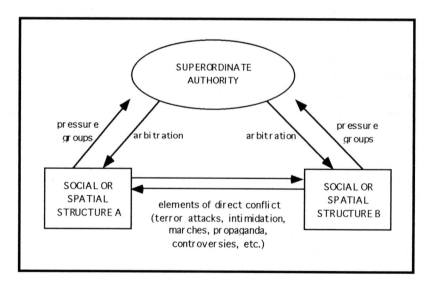

Fig. 8.3 — A model of controlled conflict under an external superordinate authority.

Conflict flared up between a fraction of the Protestant Irish élite and the British imperial élite: it reached a climax with the 1798 rebellion headed by Wolfe Tone, but it declined rapidly thereafter. Further nationalistic outbreaks were more and more the prerogative of Catholics. Even though isolated Protestants (Mitchel, Parnell) still provided Nationalist leadership, there was no mass Protestant participation to Irish nationalism after 1798.

Throughout the period of the Penal Laws, the expectations of the Irish Catholic counter-élite were met with *suppression* (total refusal). Since the late 18th century, adaptive change was introduced, beginning with a cautious *neutralization* (accepting counter-élite innovations in a very limited fashion in order to neutralize them) and giving way later, with Catholic Emancipation in 1829, to a *cooptation* (inclusion of the counter-élite into the power structure). This went so far as to the (theoretic) achievement of equal civil and political rights of Catholics and Protestants in the whole of the British Isles under Protestant English hegemony.

Persistent social and, above all, national tensions, however, led to action oriented conflict. This conflict, during and immediately after the First World War, persuaded the British imperial élite to renounce the adaptive change strategy, which it had followed since the late 18th century, and to accept a revolutionary structural change, whereby *élite replacement* and devolution of the twenty-six counties ensued. The Irish Free State was born. Unlike other colonies, however, no wholesale devolution took place.

Links of political dependence still existed between the Free State and the United Kingdom, and their very existence caused the Civil War in Eire. They were phased out only after the Second World War, when the Free State became the Republic of Ireland, and ceased to be a member of the Commonwealth. There were also links of economic dependence, and these still endure. The space of Eire and the British, however, were not contiguous: so the purely functional links which united them underwent a decided *regional divergence*, whereby the weaker national space economy, that of Eire, was fatally left behind.

The enduring political link of Northern Ireland with the United Kingdom lays further stress on the incomplete nature of the national revolution. The Northern Irish space is beyond doubt a common one, as the two groups share the same territory and the same economy. Foreign involvement, taking the form of financial support from the Irish-Americans, and arms from Middle-East terrorists, is also apparent. Adaptive change, hesitatingly attempted by the last three premiers of Northern Ireland, and later by the British authorities, after the onset of direct administration of the region in March 1972, does not seem to offer a viable solution.

The potential outcomes of the conflict will be examined later on. Unlike other cases of ethnic conflict, there is, above the struggling groups, a powerful authority: the Parliament and government of the United Kingdom. Accordingly, it is also true that the conflict takes place, at least partly, within an institutional framework which none of the struggling groups, neither the nationalists nor the loyalists, have the power (and not even the will) to destroy, but which they can only hope to influence, operating as pressure groups, armed or not. This condition is summarized in Fig. 8.3.

The three figures together provide a model aimed at explaining ethnic conflict, applicable to the case of Northern Ireland and, hopefully, to other cases as well. Alternative interpretations of the Northern Irish conflict can be grouped under the six headings: (i) paleo-Marxism, (ii) neo-Marxism, (iii) internal colonialism, (iv) historical fragmentation, (v) reductionism, and (vi) denial or apology of the domination.

(i) *Paleo-Marxism*

The paleo-Marxist utopia danced to the tune of expectations of a grand social revolution, prophesized as imminent in the most advanced countries of the time — a revolution which never materialized. Ireland has a significant place in the formulation of early Marxism, albeit in a fashion not entirely explicit. Marx expressed several times, in the many letters to his friend Engels, his views on the Irish question.

On different occasions he wrote: (i) that the aristocratic regime ruling in Ireland could have been destroyed only by British workers *after* they had overthrown "capitalism" in their own country by having a revolution; (ii) that British workers would succeed in having a revolution at home only if they had broken every link with Ireland *first*; (iii) that British "capitalism" could only be overthrown *by starting the revolution in Ireland*, as the island was the bulwark of the English aristocracy and so it provided the dominant class with means to preserve their mastery over England, since the English and Irish workers were kept divided by the "intrigues" of this class; and that was the "secret" of the "impotence" of the English working class which enabled the "capitalists" to stay in power.

The last of the three contradictory statements by Marx is particularly interesting as it contains, in a nutshell, the theory of imperialism that was to be later formulated by Lenin, who sought to explain why facts had not unfolded as foreseen by Marxian theory, and why the expected revolution had not taken place.

Of course, religion was for Marx but a dying "superstructure." On the 6th of April 1868, he wrote to Dr Kugelmann "prophesising" that, after the inevitable death of the Irish Church, the Ulster Protestant tenant would unite with the Catholic tenant of the other three Provinces of Ireland in a single movement, while up to then landlords had been able to exploit that religious hostility. Like so many other Marxian prophesies, this too is widely off the mark.

Lenin defended the Marxian interpretation — a considerably hard task, made harder by the frequent changes of opinion by Marx himself — and, in *The revolutionary proletariat and the right of nations to self-determination*, he admitted that the probability of a revolution was decreasing, but this was no proof that the Marxian "analysis" was wrong. The discrepancy between theory and reality was due, instead, to the fact that both the Irish people and the English "proletariat" had proved to be "weak."

Of course, this "explanation" explains nothing. What does it mean to say that a "class" is "weak" in the context of a theory which claims to be coherently materialistic? In terms of Marxian theory, this is a total misconception. It does not stand to its own theoretical terms of reference.

On the basis of historical materialism, one would expect a sequence of actions and reactions dialectically related to each other, in which every "class" plays its role on the basis of "dialectic" relationships with the means of production, in a purely materialistic fashion. Since this does not happen, the conclusion seems to be that *reality is wrong.*

In other words, those who do not unleash the revolution but seek rather to achieve reforms in order to improve their conditions (as the British workers did), and those who make the "wrong" revolution, the nationalistic one and not the communist one (as the Irish patriots did), are "weak," have a "wrong consciousness." In other words, *they have made what they wanted, and not what the professional revolutionaries would have liked them to do*: seize power of behalf of the "vanguard of the people," the professional revolutionaries themselves.

Social development in Ireland, as elsewhere, went its own way, irrespective of anything Marx and his followers said and did. The link between political and social issues became weaker and weaker after the agrarian reform, and in particular after the drastic Wyndham Act of 1903. Nationalistic agitation, however, was very far from subsiding, despite the fact that landlordism and the attendant social conflict were on the wane.

Lenin was adamant, on the very eve of the First World War, that the fall of the liberal government on the Irish question was a prelude to the "awakening of the working class" and the communist revolution. Though they professed to understand and judge the politics of the island, neither Marx or Lenin ever set foot in Ireland, and their ideas never took root there. The weak Irish Labour movement became, after the First World War, not less but more nationalistic.

(ii) *Neo-Marxism*

Undaunted by the failure of all Marxian "prophecies", neo-Marxists, faithful to the august example of their ideological ancestors, are still trying to show that theirs is the only theory that goes to the heart of the matter, while other theories are, at best, just a chasing after the wind, mislead by "false problems."

For example, the sociologist Wiener seeks to demonstrate that the disintegration of the Unionist Party was due to the penetration of British "big capital," represented locally, in his opinion, by the last premiers of Northern Ireland (O'Neill, Chichester-Clark, Faulkner) and by Minister Ray Bradford. Such "big capital," he maintains, threatened the small local capital and the standard of living of the Protestant skilled manpower. The conflict between reformers and conservatives within the party would thus be a false problem.

In support of his contention, Wiener gives only the fact that some important building contractors (presumably the "big capital") were interested in the construction of a ring road in Belfast. Actually very few entrepreneurs are prepared to invest in Northern Ireland, and opposition to reforms did not come so much from local small entrepreneurs but rather, in virulent fashion, from low class Protestants. Moreover — but this is probably a minor factor — there was resentment among middle class Protestants, most of them traders, against the long dominance of the landed aristocracy, still holding fast to their residual estates in spite of the agrarian reform, and still able to exercise considerable political influence.

A serious weakness of neo-Marxism in the study of the Northern Irish conflict, is the disagreement among the Marxists themselves, who thus offer a very incoherent picture of their ideological approach. It is true that non-Marxian scholars also differ widely in their interpretations, but for them it has always been normal to develop contrasting schools of thought and none of them ever claimed to possess "historical laws," and even less to be able to foresee the "end of history" or to promise Eden on earth.

It is fair to note, however, that the British political Left, namely the Labour Party, have not been particularly keen on applying "Marxist theory" to Northern Ireland, or to other political problems. It is precisely for this reason that they have been adopting a fairly pragmatic approach, and have been able to see things more clearly, thereby taking a far more realistic and practical stance on the Northern Irish problem than the Tory establishment.

(iii) *Internal colonialism*

The old internal colonialism theory by Hechter is rather unsatisfactory. First of all, Ireland is a quite different nation from England: an island, entirely separated, geographically, from other nations. Why then speak of "internal" colonialism? That of Ireland, instead, is a case of entirely external colonialism.

Confronted by Eire's devolution, Hechter is of course obliged to take stock of the peculiarity of the case, which makes Ireland so different historically and politically from all other regions of the so-called "Celtic periphery." Such devolution, according to him, is due to the particular pattern of dependent development which emerged in Ireland between 1846 and 1921, that is to say to the formation of a regional agrarian economy with a comparatively high capital density which did not lead to substantial inter-regional organizational affiliations, as emerged in the case of the "highly restricted" industrial development of Wales and Scotland.

It should be made clear from the outset that it is utterly wrong to regard the industrial development of Scotland and Wales as "highly restricted", as Hechter does. The two countries, instead, took part in the Industrial Revolution in full, together with North-East Ulster and unlike all the rest of Ireland. Secondly, Hechter does not go deeply enough into the cultural and political nature of Ireland's dependent development and the links it has with Eire's devolution. To take as a starting point for this devolution, the year 1846, is arbitrary and evinces a poor historical perspective, at least for two reasons: Hechter neglects the revolutionary role of the Irish who had emigrated to the USA, and he hurriedly pushes aside all the previous stages of the conflict.

This is linked to a refusal to take into account, among the causes of the devolution, the peculiar (and often inhuman) treatment meted out to the Irish, and the cultural differences between Ireland and England. The basic mistake is an extreme economicism, a negative heritage of Marxian origin. As usual, *he who sees only the economy does not understand even that.* The Marxian dogma picturing religion as a "superstructure," dependent upon the economic "structure," is not openly stated by Hechter, but insidiously assumed throughout.

Starting from these questionable assumptions, Hechter ignores the role of religion in the conflict process and its cultural impact on the nature of the élite. Also, he does not come to grips at all with the concept of the degree of closure of an élite, strongly linked in this case with religious affiliation.

To ignore the role of religion makes it exceedingly difficult to understand why, in spite of the conflicts which erupted at different times against English dominance, Scotland, Wales and Northern Ireland itself have not (or not yet) undergone devolution from the United Kingdom, and why the various parts of the "Celtic periphery" did not make a common cause against the English.

Moreover, why did the Scotch-Irish Protestants side with the Anglo-Irish Protestants rather than with the Irish Catholics, over the Penal Laws which affected them too? Also the fact that farm rationalization and specialization in animal husbandry took place in the remainder of the British Isles begs the question why such economic trends had a very different effect elsewhere. In Wales and Scotland, the English did not persecute religion, which was overwhelmingly Protestant, though not Anglican: hate was concentrated on Catholics, and therefore on the Irish, regarded as the evil agents of "Popery." Likewise, neither in Wales or in Scotland did the English seize the land of the natives *en masse*. Only Irish land was seized and doled out to adventurers and soldiers.

But why did the English refrain from seizing the land in Scotland and Wales and not in Ireland? Is it not because in Ireland Catholic landlords had to be made politically impotent since Catholicism inspired more diffidence than Scotch and Welsh dissent? Even the land question appears therefore, to some extent, to be linked to religious antagonism.

The rights to religion and to the ownership of land, grossly violated in Ireland and largely respected in Scotland and Wales, provide a more plausible explanation as to why Scotsmen and Welshmen, though harshly discriminated against at times and ready to oppose English dominance, did not (yet?) go the way of the Irish.

On the other hand, though steering clear from economicism, one has to avoid the opposite mistake of neglecting the economic factors: having taken part in the Industrial Revolution most certainly contributed substantially to lead Wales, as well as Lowland Scotland, towards integration within a unified British space economy.

(iv) *Historical fragmentation*

Historical fragmentation is, at least in part, the upshot of a professional deformation of historians. By becoming specialized in the study of a given period, they often tend to stress its peculiarities and discontinuities, to the detriment of continuities in structures and processes. Not all historians, evidently, have benefited from Braudel's lesson on the *long durée*.

In the case of Ireland, moreover, revisionist theses which deny continuity to the conflict can be of use in undermining extremist views, and have therefore a laudable, though unexpressed, political purpose.

It would be futile, in fact, to deny the existence of paradoxes, discontinuities and ambiguities in Irish history, as in any other national history. One has only to consider the first plantation ordered in 1558 by the Catholic Queen Mary, or the loyalty to the Crown by Catholic militia units in the struggle against Protestant insurgents in Ulster in 1798. On the other hand, one cannot deny a basic continuity of social processes fuelled by the persistence of a structural conflict situation caused by an alien domination.

(v) *Reductionism*

The reductionism of Irish motivations is often implicit in revisionist theses, as they tend to equate the ideal drive of Republicans with "mere nationalism," ignoring the universal longing for freedom, which doubtless is a powerful factor of ideal continuity.

Some authors tend to ignore conflict elements. Cullen, for example, in his economic history of Ireland, attacks the traditional view according to which Irish development was constrained by dependence, and lays stress on private and business documents, rather than upon political and administrative sources. This approach can be satisfactory to those wishing to highlight the history of a particular family, an estate, a village, not to those who would like to understand why so much blood has been running in the streets of Northern Ireland, and of the rest of Ireland as well.

Some scholars seem to resent history as such. Foster speaks of "imprisoning historical perspectives." He notes with disapproval that Ireland's history has been told so often as a sequel of wrongs, persecutions and disasters, and has been used to justify republican terrorism and civil war. This is undeniable. But why has it been told that way? Because *it is* a sequel of wrongs, persecutions and disasters. Unfortunately, anything that happened in the past, is still at our elbow in the present, and we will carry it with us into the future. It cannot go away, because it lives with us, in the lasting consequences it inevitably has. If "historical perspectives" are "imprisoning," it cannot be helped.

Moreover, *it is very easy for the descendants of the invaders, and of their local abettors, to suggest sweeping history aside.* The Turks have long forgotten that their ancestors slaughtered the Armenians. To the Armenian mind, the harrowing thought of the Turkish massacres is constantly present.

Finally, Irish history is too little known. The man in the street in Britain does not understand the conflict in Northern Ireland precisely because he does not know (and perhaps does not wish to know) Irish history. In this way, as pointed out in Chapter 4, it is very easy for the British media, not particularly sympathetic with Irish republicanism, to make the conflict utterly unintelligible to the British public.

Reductionism may also take the form of laying particular stress on the people "in between" (Boyle & Hadden): the moderates who, refusing to take sides, may become the means whereby peace could be achieved. Unfortunately, they have proved, so far, to be utterly powerless. As stressed before (Chapter 4), their only chance to take the initiative and rule Northern Ireland in a moderate fashion, by means of a democratic Executive came in 1974, after the Sunningdale agreement. Their Executive was speedily quashed by Paisley's extremists, while the London government did not budge. So far, the people "in between" have not had any further occasion to act in a reasonably effective way, neither will they have any, unless the British government stops courting the "agreement" of loyalist extremists.

(vi) *Denial or apology of the domination*

Revisionist tendencies in general tend to skip the theme of Protestant political and social domination in Ireland, or deny that it had any real relevance. From soft-pedalling to outright apology, the step is but a short one.

Books like J.C. Beckett's *The Anglo-Irish tradition*, argue that the "rejection" of the Irish Protestant tradition by the Catholic majority has been highly damaging to Ireland: had the "maturity, wisdom and broadness of outlook" of the Protestants not been rejected, "even a united Ireland could have emerged". Of course, had the Ascendancy been left in power, Ireland would doubtless still be united today, as an obedient colony.

The apology of the domination may be even more bare-faced. Someone, like Jackson, has laid the blame of the present situation in Northern Ireland squarely upon the Catholics, who "attract discrimination" by refusing to accept assimilation, in other words because they stubbornly persist in being Catholic and do not make the Protestant-Unionist-Loyalist views their own. This frenzied attitude is not even a mere apology of domination, but of persecution. Jackson denies everything: the Stormont regime never tried to influence economic development to favour the East of Northern Ireland; the struggle is merely a matter of nationalism and has nothing to do with religion; as if religion had not played a decisive role in shaping nationality itself, in Ireland as anywhere else.

It is plain, as many authors have stressed, that *Irish nationalism arose precisely as a response to the stubborn attempt of the Protestant colonial minority, planted and supported by the British, to preserve at all cost, its religious, political, social, and economic privileges, at the expense of an impoverished majority full of resentment.* British concessions (adaptive change) always came too late and nationalism became more and more rigid, leading to repeated outbursts of violence. The whole situation was, and is, socially pathological, in the sense envisaged by the imbalance multiplier (Fig. 8.1), and such as to fuel a sharp and lengthy conflict, and to revive the same conflict even after it had subsided for some time.

One final point is that Irish unity is still a controversial issue because Irishmen labour under several disadvantages, as we have seen. They are too weak: had they won earlier, they would not be so tired of such a long-drawn struggle, as they themselves are. Had they won earlier, Northern Ireland would have developed, socially, economically, and politically, on quite a different path: for instance it would not harbour such contrasting political extremisms, so much so that assimilation by the Republic has become more and more difficult, and it is unclear now whether the Irishmen of the Republic actually want reunification to take place.

But the main disadvantage is, paradoxically, the colour of the Irishmen's skin. We have to take stock of the fact that, if the rulers and their subjects are of the same colour, they do not hit the headlines in the same way as when the rulers are white and the "underdogs" black. Had the Irish been black, or at least "coloured," they would have been submerged by the world's solidarity, they would have commanded universal support. Britain would have been condemned, day in day out in the United Nations, precisely as South Africa was. Unionists would have been regarded in their true light: *a settler community to be respected, certainly, but not to be allowed to stand in the way of decolonization.* Thus a united Ireland would have probably been achieved long ago.

Since Irishmen are white, there is no "politically correct" capital to be made out of them, so that, in their particular case, even denials or apologies of the colonial domination can still be concocted with impunity.

A further useful comparison can be added here. Something similar to the break-up of Ireland in 1921 might have been repeated in the early Sixties in another colonial conflict. Towards the end of the Algerian war, a queer idea was mooted in France: upon the granting of independence to Algeria, a

European enclave was to be built, by clipping out a chunk of territory around Oran, where the Europeans were actually in the majority. Luckily, the plan was not taken seriously. It would have just brought about more bloodshed, and Oran would have had to be given up just the same. The Algerians were stronger and had more international backing. The Irish were weaker and largely ignored by world opinion, except, to some extent, by the Irish Americans. That is why the French-Algerian war is over (Algeria's present problems stem from quite different sources), and Northern Ireland is still "a problem."

CHAPTER 9

BEYOND NORTHERN IRELAND: THE PRACTICAL PROSPECTS

What are the prospects for a solution? To answer this question one must bear in mind: (i) the internal conditions of Eire and how the Republic can help in achieving a solution, (ii) the internal conditions of Northern Ireland, (iii) the political and economic situation of Great Britain and therefore, which way the London government can move, and wishes to move, in the quest for a solution. In view of the weakness of the Republic in comparison with the United Kingdom, the potential contribution of the former is doubtless the least important.

(i) *The weakness of Eire*

Eire is regarded by the Protestants of Northern Ireland as a Catholic theocracy. Nothing could be farther from the truth, and this misconception aptly highlights to what extent the Protestant-Unionists can be out of touch with reality.

The declining influence of the Church hierarchy; the harsh conflict between the Church and government on the public health policy between the late Forties and the early Fifties; the downgrading of the censure law, the significant changes that have taken place in Catholic social thought; the election in November 1990 of a President of the Republic who is a feminist supporter of abortion and divorce; the growing secularization of Eire witnessed recently by the referendum in November 1995 which has introduced divorce; all this has been lost on the Northern Irish Protestants.

In Eire, the attitude towards Britain is very ambiguous: an inextricable mixture of admiration and resentment, emulation, and spirit of revenge. The situation in the North tends to rekindle traditional negative attitudes towards the great neighbour, but the Republic's strength, both internally and in its international relations, is sapped by its small size and by the crippling

181

economic crisis. This accounts for the inability of Eire to exercise a credible political pressure to obtain the six Counties.

To this must be added the massive imbalance between the welfare systems in the two Irelands, an imbalance that contributed a great deal to strengthening the partition and the dependence of Northern Ireland upon Great Britain. The British welfare system was introduced in Northern Ireland largely at the expense of the British tax payer. Should the six Counties be united with Eire, the finances of the latter would be put under an intolerable strain to keep such a system going; besides they would also have to face heightened social demands in the remainder of the country.

(ii) *Economic weakness and political division in Northern Ireland*

Northern Ireland, on the other hand, is not economically self-supporting. Should it become independent, it would be confronted by the dilemma of whether to downgrade welfare expenses dramatically or to increase taxes in a likewise drastic fashion: in view of the extremely high unemployment rates, both solutions would be equally catastrophic.

Most wage and labour condition improvements in Northern Ireland are but the result of struggles by the British trade unions. The introduction of such innovations into the region has been a mere matter of passive transfer. The eminent feature of the Northern Irish trade union activity, dominated by Protestants, has been, rather, anti-Catholic discrimination. As stressed by Boal and Douglas, to discriminate in hiring and promotion policies against a weaker minority requires less effort and gives more certain results than a common action against a powerful dominant class.

However, the times of unchallenged discrimination and undisturbed dominance will hardly return. The conflict triggered by the brutal aggressions to peaceful marchers in the late Sixties appears irreversible, except through the onset of drastic structural change; and the growing Catholic community will not be satisfied with anything less than the union of the six Counties with Eire.

The legitimation of Northern Ireland hinges upon a simple question: is the Northern Irish Protestant community a true majority? It is, in the six Counties arbitrarily snatched from Ireland. But it is actually a minority both in Ireland as a whole and in the United Kingdom. No longer does it wield political power upon the region. For example, Protestant extremists in jail demand in vain, to be segregated from the Catholic ones, of whom they are afraid.

Devoid of a secure territorial base and having lost its former unchallenged power, the Protestant-Unionist community behaves, in general, not as a majority but as a minority under siege, frightened by every change, holding on to Paisley's slogan: "No surrender." On the other hand, the Catholic-Republican community shows greater reliance and determination, with a feeling of belonging to the Irish majority.

It is true, on the other had that, despite its small size, Northern Ireland contains a great variety of situations. As stressed by Whyte, there are areas such as the southern part of Co. Armagh, which are violently nationalist and republican and others such as northern Co. Down which are massively unionist; and still other areas, like the town of the Ards peninsula (Co. Down) studied by Buckley under the faked name of "Upper Tullagh," where unionists and nationalists are in intimate but peaceful contact. However, "Upper Tullagh" is three quarters Catholic: would social relations be so relaxed and devoid of discrimination should the town be dominated by Protestants such as the followers of Ian Paisley?

If all Northern Ireland, argues Whyte, were like the former of the three areas, there would have never been any partition. If it were like the second, the IRA would find no support, and the whole region would indisputably be a part of the United Kingdom. If it were like the third area, it would be possible to find "some kind of compromise." Since the region contains all three areas and many others besides, with a variety of intermediate situations, the search for a compromise could aim at a multiplicity of solutions for the different areas.

It is a very feeble idea: by watching the trees, Whyte loses sight of the forest. Perhaps what he had in mind was some sort of cantonal solution. Yet Northern Ireland is an urbanized and economically integrated society. It is no Switzerland, has no valleys, culturally and physically distinct. And still the main problem would remain: what kind of institutional framework for Northern Ireland as a whole? Even the Swiss had to resort to a common federal arrangement.

Whyte's conclusion, that he would be satisfied with *any solution* of the crisis that might bring peace, is highly disturbing and utterly unacceptable. Peace without freedom and justice is but a mockery. The very concept of "any solution" is based upon an appalling moral indifference, and it rests on the assumption that different solutions are possible, whereas there is only one.

The one solution is a united Ireland. Ireland should not have been divided in the first place. There was no need for a crystal ball to realize, at the time of partition, that such a solution would only bring about endless trouble. The IRA will not have peace until Ireland is united, and therefore, the only way to peace, the only realistic way, is to allow Ireland to become united.

Would this mean yielding to the IRA violence? But how many political solutions have been brought about by violence? The American Revolution did not take place by throwing flowers; nor was the end of apartheid achieved simply by gentlemen's talks; nor was Kenya's independence obtained merely by friendly persuasion.

Yet no one questions the legitimacy of George Washington's America, or of Nelson Mandela's South Africa, or of Kenyatta's Kenya. To this might be answered that the IRA do not represent the Irish people. This is undeniable. However, no revolutionary élite ever represented the whole people. In the Thirteen American Colonies, there were many loyalists: some were persecuted and killed, many emigrated after the Revolution and founded Ontario. Even in South Africa, many blacks supported the regime and were afraid of the African National Congress and the Pan-African Congress. And certainly, a mere small minority of Kenyans turned *mau-mau*, whereas many Kenyans at the time were afraid of change and would have preferred the British to stay. Yet a solution was achieved, in these and many other conflicts, when the ruling élite decided that to keep fighting was not worthwhile.

If the problem is merely to end the conflict in Northern Ireland for good, it is useless to question the morality of dealing with "terrorists" (and thereby to some extent legitimizing them). After all, the covert units of the British intelligence (as any similar body anywhere else in the world) are not waging a "clean" war either, as Dillon's fundamental books on the Northern Irish conflict aptly demonstrate. And even "dirtier" is the behaviour of the loyalist paramilitaries.

Nor is there any question of "yielding" to the terrorist threat. Kenya was given up to the nationalists and ceased to be "a problem." The time was ripe. Now, time may be ripe for a united Ireland. And, obviously, uniting Ireland is by no means tantamount to delivering Northern Ireland into the hands of the IRA. It can be delivered into the hands of the legitimate government of Eire, where the organization is outlawed.

It is true that the IRA are against the government of the Republic too but, should unity be achieved, their bugle calls would no longer attract recruits. Uniting Ireland would be the surest way to turn IRA men into helpless misfits, who indeed will have serious readjustment problems in their very homeland. Ireland, in any case, would remain within the cultural sphere of England and the English-speaking world, like the whole of Europe and the whole of the planet, by the way.

The need to put an end to a senseless struggle goes even beyond the moral question of whether or not the Irish have the right to be united. Even if a united Ireland were a most unfair option, the only *practical* way to peace would still be a united Ireland.

Because, after a final — and most probably dreadful — outburst of violence, there would be no further reason to fight. *The Unionists can lose only once*, and then they will have to come to terms with their defeat, or emigrate. *The Republicans, if defeated this time, will start the fight all over again*, at some later stage, even though outlawed in the Republic itself. In a Northern Ireland linked to the United Kingdom, the struggle may have no end. And the struggle will spill over into England again, affecting innocents, as has already happened, as always happens in any war, no matter whether it be open war, guerrilla war, terrorist war or gang war.

Though exceedingly rich in information, Whyte's celebrated book does not go very deep. Clarity of thought gets lost in a maze of statements and counter-statements, whereas clarity of thought is, by definition, the ability to discriminate between what is essential and what is not, between truth and red herrings.

The misconception eating up the very roots of Whyte's otherwise informative work is an obsessive relativism. He expounds all the opinions and criticizes them rather superficially. Finally, he fails to comes to any intelligible conclusion. He is not alone. The very concept of truth seems to suffer a disastrous crisis in the modern world. In this regard, Whyte's book is an admirable specimen of the present academic fashions[*].

On a more practical vein, some politologists (Lijphart, Rose) have

---

[*] At one time, academics were bent on discussing supreme truths, the meaning of relevant historical events, the legitimacy of the king's decisions. Universities formed leaders. They formed men. Nowadays they form professionals and bureaucrats. The true object of education has been forgotten. Overspecialization is leading to a new kind of illiteracy, while the supreme purpose of man is the quest for truth and good. These concepts have recently been authoritatively expressed by the Chancellor of the Universidad Complutense of Madrid, Gustavo Villapalos Sala, in an interview to the *Giornale Nuovo* (15th May 1995) of Milan.

Undeniably, science cannot be expected to achieve truth in an absolute sense, for this is not what science is for. But science is not all. If truth were unattainable, human dignity would be lost. The denial of the possibility to achieve truth evinces only the infirmity of man, who is afraid of truth. *But this retreat into an exasperate relativism is self-defeating.* Relativizers are seldom sincere: they use relativism as a mere tool to relativize, and thus be able to reject, what others say. However, by their very denial of the possibility to know the truth, the dominant academic tribe of the relativizers fall into an inescapable contradiction: why should their opinions be "true," and therefore acceptable, if truth cannot be known?

discussed the partition of Northern Ireland, but only to reject it, in view of the extreme difficulty of tracing acceptable frontiers. In spite of local segregation, Catholics and Protestants live in close economic inter-dependence on the same territory. Northern Ireland is devoid of natural or ethnic borders, and also of historically justifiable borders.

At least eleven of its twenty-six districts have Catholic majorities and, if allowed to choose, would most probably opt for the Republic, leaving the other districts with an overall area which would be nearly halved, and an even more absurd political border. Already at the time of partition, two of the six Counties had Catholic majorities: their inclusion into the new statelet amounted to mere usurpation. Even the name is "inappropriate," to say the least: the northernmost point in Ireland, Malin Head, is in Co. Donegal, and belongs therefore to the Republic. And obviously, the name "Ulster," which the Unionists regularly prefer, is utterly wrong: the province was partitioned too, and three of its nine Counties were left to Eire at partition.

"Northern Ireland" is the official tag of something which has not even a proper name. If, in order to ward off any possible mental association with Ireland, the Protestant-Loyalist-Unionists do not even like that official tag of their statelet, but prefer to call it "Ulster," if the very name "Ireland" is so unpalatable to them, what are these people doing there? At least the South African Boers liked Africa so much as to call themselves Afrikaners, which in the original Dutch means "Africans." And eventually they accepted other Africans into a common citizenship.

It is highly doubtful whether the Protestants of Northern Ireland can be regarded as a nation: they themselves, usually appear to be certain of one thing only: they do not wish to be a part of a united Ireland, while positive identification is weak, except for an overall attachment to "Britishness" and to "Ulster." "No" seems to be the word they use more often in political life.

However, the conflict itself has stimulated a twofold process of national formation. On the one hand, Catholics have strengthened their sense of identity, which they feel is linked to the Celtic cultural heritage. This feeling is likely to come closer to the truth in Northern Ireland than elsewhere, the reason being that Anglo-Norman power and influence were rather weak in the area, and very little inflow of pre-Reformation Catholic "Old English" took place. In other parts of Ireland, and especially Leinster, where "Old English" immigration was considerable, Catholics have a definitely more mixed ancestry.

Protestants have tended to hark back to a far more undefined cultural background, linked both to the colonial tradition, and to the English and Scottish national heritages, preserved, to some extent, by the allegiance, respectively, to the Anglican or the Presbyterian Churches, though lately several minor Protestant sects have been growing fairly rapidly. But a disturbing cultural insecurity lingers on. A longing for a separate "identity of Ulster" is felt by some, as a cultural coccoon against Irish "encroachment."

Significantly, there have been some (Adamson, Hall) who, in pseudo-scientific books entitled *Cruthin: the ancient kindred, The identity of Ulster, Ulster: the hidden history,* have sought to satisfy this rather romantic wish. Ulstermen (presumably Ulsterwomen too) are said to be the offspring of prehistoric populations of whom, conveniently, very little is known. Nothing to do, therefore, with the alien Irish, who just by chance happened to be hanging around in "Ulster."

Others (Buckley, Gailey), with greater scientific rigour, have sought to show that there are no separate Catholic and Protestant traditions; but that, in dialects, agricultural practices, housing types, manufacture of crockery, cooking, family life, and private morality, cultural differences are more a matter of regional (Robinson) or class variation than Church allegiance. No one would take issue with this, but there remains the fact that using the same kind of cooking pots unfortunately does not prevent extremists from both sides from planting bombs and slaughtering each other, or, more frequently, from slaughtering innocents. Beyond cultural similarities, the conflict exists as an inescapable reality, and has well defined historical roots, reaching far into the past, although it might be stilled by a ceasefire.

Cultural similarities with their Catholic fellow citizens do not prevent the Northern Irish Protestant-Loyalist-Unionists to play, politically, a role very similar to that of the French settlers in Algeria or of the Rhodesians who were once privileged minorities, which, as times were changing, lost their mother country support. If the Northern Irish Protestants are not a nation, they cannot possess a right to self-determination. Their uncontrolled regime has proved inept in providing a modicum of justice and order. No doubt, restoring the autonomy of Northern Ireland would be an exceedingly risky operation.

There is in fact no decline, among the Protestant lower classes, especially in urban settings, of the discriminatory attitude towards Catholics. The rejection, not only among the lower classes, but also among middle and upper class Protestants, of the idea of belonging one day to a united Ireland, remains unabated, no matter whether they deem this development unavoidable. After the prorogation of Stormont, even Britain has become unpopular. In Protestant working class areas, the Ulster flag is often displayed nowadays, rather than the Union Jack.

A proposal was mooted by prominent representatives of Protestant extremism, in particular William Craig, that Northern Ireland declare unilateral independence, like Rhodesia, and that the financial weakness of the new State be made up by replacing the financial support from London (which, in all probability, would be immediately phased out) with Soviet grants in exchange for the concession of one or more military bases. In cold war conditions, this would have enabled the Soviets to threaten the North Atlantic routes, thereby checkmating NATO air and naval forces and starting an encircling maneuvre of the whole of Western Europe. The changed international situation makes such a scenario utterly ludicrous, but the fact that a proposal of this kind has been uttered is highly suggestive of the nature of Protestant extremism.

On the other hand, Catholic extremists, even by uniting their three components (PIRA, OIRA and INLA) would have no hope of winning by their forces alone. All they can do is to bring pressure to bear upon the London government through terrorism or negotiation, in order to induce it to opt for a political solution which might, in due course, bring about a united Ireland. The possibility that a solution of any description might arise from within Northern Ireland, instead, must be decidedly excluded, the warring factions being too irreconcilable.

If the Republic is too weak and Northern Ireland too divided, only one possible source of a solution is left, namely Great Britain.

(iii) Conflict resolution from Great Britain?

It must be pointed out, first of all, that the unrest in Northern Ireland is not really relevant to the British authorities and the public, who are not particularly concerned when "things happen up there." This readiness to countenance "an acceptable level of violence" — whatever that may mean — provided it takes place elsewhere, has been well realized by the IRA leadership, who accordingly decided to take terrorism into "enemy" territory: to London and England at large, though not to Wales and Scotland, which are regarded as kindred Celtic lands.

Attention must then be turned towards the United Kingdom in order to evaluate the possible outcomes of the conflict. For generations the political link with Britain has been regarded by the Irish Protestants first, and then by those of Northern Ireland, as the best protector of their dominance and privileges. Lately, however, the Protestant Northern Irish and British public opinion have become increasingly estranged. The demise of the imperial idea and the growing secularisation of society have made the prejudices still dominant in Northern Ireland appear more and more obsolete to the British.

Already the Tory Party, even under an intransigent leadership such as that of Mrs Thatcher, has been led to concessions unpalatable to Northern Irish Protestants. What will happen when power in the United Kingdom will revert to the Labour Party — which sooner or later must be the case in the normal democratic game?[*]

It is no coincidence if Irish immigrants in Britain tend to support Labour. Through Labour and the trade unions, they make up a power block that is considerably stronger than their numbers would warrant. Their vote is decisive, in fact, for about thirty seats in the Commons. The Labour Party has often sharply criticised Tory intransigence in dealing with the Northern Ireland problem. This opposition was spelt out in 1987 in a white paper entitled "New rights, new prosperity and new hope for Northern Ireland," which contains a specific promise to act in favour of a united Ireland. Protestant-Unionists will not be allowed to veto any political development. In a later document of 1988, the party confirmed this attitude. But are politicians to be trusted?

Early in 1995, and perhaps earlier, Mo Mowlam, the Shadow Northern Ireland Secretary, was developing links with the Unionists, while in the Unionist party serious consideration was being given to a cooperation with the Labour Party, Since the demise of the rugged Michael Foot, however, the Labour Party has doubtless moved a great deal to the right. This does not bode well for a firm stance by them on the issue of Northern Ireland. On the other hand, if present trends continue, the Labour Party, after the next elections, will enjoy such a majority in Parliament as to be able to do without the Unionist support, and this may create the conditions whereby the Unionists may really be in no position to veto "any development."

In one way or another, the idea that Irish unity is, sooner or later, unavoidable, is making inroads even amongst the staunchest Protestant-Unionists-Loyalists in Northern Ireland, after being for a long time, a persuasion of their Catholic-Nationalist-Republican fellow citizens alone. As we have seen, it is possible that the higher birthrate of the latter will enable them to achieve the majority early next century. But in any case, leaving aside dubious attempts to formulate accurate forecasts, the psychological effect of a continuing growth differential may be strongly felt. Moreover, emigrants are nowadays mainly young well educated Protestants who fail to find satisfactory jobs. Even if the two groups become "stable" at an

---

[*] One might be tempted to add that, in view of the Tory disaster in the 1995 local elections, it is probable that Labour will rule the United Kingdom in the very near future, possibly after the 1997 political elections.

approximately equal size, it might be exceedingly difficult to keep Northern Ireland indefinitely within the United Kingdom.

According to an opinion survey carried out by the *Guardian* in November 1993, 40% of the British are in favour of a devolution of Northern Ireland, 16% favour joint responsibility of the London and Dublin governments for running the region, 26% of the sample have no set opinion on the matter, while a mere 18% wish Northern Ireland to remain a part of the United Kingdom. Significantly, 59% of the sample are in favour of direct negotiations with the IRA.

For, what use do the British have for the six Counties?

Northern Ireland is a colony. Colonialism is dead. Therefore, the Northern Irish colony is but a leftover of a past age and does not make sense today. A simple sillogism against which it is difficult to find a rebuttal.

Moreover, Northern Ireland is a security risk and an international liability. And further, it is but a financial burden to the British taxpayer. Finally, it is utterly useless for the defence of Britain.

During the Second World War, bases in Northern Ireland helped protecting the Atlantic routes from North America, while Eire remained neutral, probably because Premier Eamon De Valera deemed a German victory more likely and thought that a united Ireland might then be achieved without spilling Irish blood. During the Cold War, the same bases could have been potentially useful in the same way, although a more rational solution would have been to allow Ireland to become united in exchange for its joining the NATO alliance. Eire, instead, consistently refused to join NATO, precisely because that would have implied the acceptance of existing borders.

Nowadays, however, both the changed strategic situation and the highly sophisticated military hardware make military bases in Northern Ireland an unnecessary luxury for Britain. Moreover, Ireland will always be politically and culturally linked to the USA, so that, in order to ensure having Eire on their side in case of a crisis, the United Kingdom will only have to keep its present close ties with Washington.

Irish-American communities have been a decisive factor in the past. But what of today? The USA are steeped in both traditions: the Celtic and the Anglo-Saxon, and in many other cultural traditions as well. Even among the Irish-Americans, many are no longer particularly interested in the issue of Irish unity and have no clear ideas on the matter. The overall interest of the United States, the only superpower on Earth, tends to peace-keeping (or *status quo* keeping), to avert dangerous crises and protect their own giant economic interests: this political philosophy is basically the same whether applied to the Persian Gulf, Bosnia, or Ireland.

The intervention by President Bill Clinton late in November 1995, not in favour of the Irish side, but, in a more detached fashion, merely to "revive" the stranded "peace process," is in keeping with this philosophy. An international committee, chaired by the American Senator George Mitchell, was allotted the task of discussing the "practical" issues, i.e. the technicalities, of the political negotiations, which in reality never got anywhere.

Despite the assurances from the Tory establishment, the Northern Irish Protestants seem to be rather discouraged. A current bitter jibe among them is: "the British will sell Northern Ireland."

If it came to that, however, even the best contrived institutional devices to insure a painless transition (interstate consultative assemblies, regional federative structures with increasingly autonomous powers, and so forth), will hardly persuade the Protestant-Loyalist-Unionists to accept annexation to the Republic without a struggle. The Unionist paramilitaries would unleash the customary spate of indiscriminate murders and terror attacks, most probably hitting not only the Catholic areas of Northern Ireland, but also the territory of Eire, as they have done in the past. The IRA would hit back.

The Republic would hardly be in a condition to take over. It is doubtful that the Irish armed forces would prove powerful enough to hold the whole of Northern Ireland against the Loyalists, but they could control large areas with a Catholic majority, such as Counties Fermanagh and Tyrone, southern Armagh, the towns of Derry and Newry.

Protestant Ireland might, therefore, shrink a great deal. However, even the most severe territorial sacrifices would not bring about a solution, since its territory will in any case still hold extensive Catholic-Nationalist areas, including West Belfast, a stronghold of the IRA. Territorial losses would make Northern Ireland increasingly weaker economically, until a wholesale unity may become inescapable.

The more extremist supporters of the lost Loyalist-Unionist cause might feel forced to emigrate, but nothing would prevent the Protestants from staying and becoming citizens of the Republic, with full equality of rights, as the Protestants in Eire, who, far from suffering any disability, are in fact a privileged minority. Moreover, nothing would prevent the Irish Protestants in a united Ireland from keeping their cultural links with Britain.

(iv) *An unlikely scenario, but not utterly impossible*

A further, and quite revolutionary scenario could also be envisaged, though it seems rather unlikely, at least in the short term.

Let us consider the following elements:

(i) the growing support, in Scotland and Wales, to nationalist parties seeking a broad autonomy (Home Rule?) or outright independence;

(ii) the fact that the United Kingdom is regarded by nationalists of the Celtic periphery as an imperial leftover (which in fact is not very far from the historical truth);

(iii) the decreasing prestige of the Royal family (the chief symbol of political unity in the United Kingdom), due to the well known occurrences widely publicized by the media, and especially by the popular tabloids;

(iv) the fact that England's comparatively prosperous economy (especially the South-East) subsidizes the other parts of the UK, which may, in future, lead perhaps to the rise of a new *English*, as distinct from *British*, nationalism, demanding England's independence.

Will all these different processes develop and lead to the dissolution of the United Kingdom? In this case Irish unity would follow as an inescapable corollary.

Even a partial realization of the potential factors for a change listed above, without leading to the demise of the UK, might have strong repercussions on Northern Ireland. For instance, a heightened pressure by the Scottish and Welsh nationalist parties might bring about autonomy for the two nations. At that stage, the London government would be obliged to reconsider the constitutional position of Northern Ireland, and would then be confronted by the dilemma: restore the region's autonomy? But how? By allowing the loyalists to take over again or by trying to protect the Catholics? And in what way? Any kind of solution would be fraught with serious risks and uncertainties.

(v) *A glimpse of a united Ireland*

If a united Ireland appears, in the foreseeable future, extremely likely, if not inevitable, it would be foolhardy to understimate the immense practical difficulties, economic and political, confronting the new State.

A future united Ireland will need external help for a considerable time. Within the European Community, one could envisage an "Ireland Special Fund," whose task would be to help the country become economically united and self-supporting. Cooperative agricultural development, transport infrastructure, high tech industry, offshore economics, educational facilities, scientific and technological research, high level services, environmental protection, tourist development and of course job creation, should be the main priorities. The American Irish, or at least those of them who still feel attached to their Irish roots, could also play a significant role, similar to that of the American Jews in helping Israel.

An essential part of any development policy for a united Ireland must be the drive to ensure equality of opportunity to people and regions. In this regard, special attention will have to be paid to the lagging Catholic areas of the present Northern Ireland.

The peripheral insular location need not be an obstacle to a healthy development. Many factors (priority of human over natural resources, footlose industry, computer networks) have radically changed the traditional development paradigm based upon centrality and a powerful urban polarization. The very meaning of a term such as "peripherality" has accordingy changed. A united Ireland, with a larger internal market, a better integrated space economy and a highly educated workforce, could be in a pole position to exploit the opportunities offered by an ever changing world economy.

The unity and welfare of Ireland must be a common responsibility for all of us as Westerners, and in particular as Western Europeans. Western Europe is one of the most advanced and powerful areas in the whole planet but, to be at all credible in any attempt to find a solution to conflicts elsewhere in the world, it must first show that it can solve its own conflicts.

REFERENCES

Rather than trouble the reader with frequent references, either in the text or in footnotes, I have preferred to indicate just a very few of the most important authors in the text by name only, without a formal reference. Hereby are listed all the books, articles, and sources of documents that have been consulted in the preparation of this book. In particular, the theoretical approach that has been used is based on a general theory of development that I published in the *Discussion Papers* of the University of Southampton. This theory, of course, did not spring out in armoury and full attire by itself. I am indebted for many of its concepts to several important contributions by geographers and other scholars listed in the first section of this reference list, among whom Bird, Friedmann, Pinder and Popper have been particularly useful.

General and methodological essays

ARGERI D. (1979) *La dialettica dissacrata*, Milano, Sugarco

ARON R. (1951) «The Leninist myth of imperialism», *Partisan Review*, 18: 646-662

BESANÇON A. (1981) *Anatomie d'un spectre*, Paris, Calmann-Lévy

BIAGINI E. (1979) «A conflict theory of development» (preprint)

BIAGINI E. (1981) «A general theory of development», *Discussion Papers*, Department of Geography, University of Southampton, No. 10

BIAGINI E. (1981) «Analisi critica dei contributi interdisciplinari per una teoria dello sviluppo», *Bollettino della Società Geografica Italiana*, X, X: 423-462

BIAGINI E. (1982) «Stadi di sviluppo: una formulazione teoretica», *Rivista Geografica Italiana*, LXXXIX, 2: 332-346

BIAGINI E. (1983a) «Potere e manodopera a buon mercato nello sviluppo della Colonia del Capo, 1652-1795», *Bollettino della Società Geografica Italiana*, X, XII: 55-74

BIAGINI E. (1983b) «Sviluppo a "frontiera discontinua" e processo conflittuale nel Sud Africa», *Rivista Geografica Italiana*, LXXXX, 2: 193-249

BIAGINI E. (1984) *Sud Africa al bivio: sviluppo e conflitto*, Milano, Angeli

BIAGINI E. (1992a) *Pianificazione territoriale in Occidente*, Bologna, Pàtron, 3rd edn.

BIAGINI E. (1992b) «Inghilterra, Irlanda e Nord America: relazioni marittime e conflitti Centro-Periferia dopo la scoperta di Colombo», *26th Congresso Geografico Italiano*, Genoa

BIAGINI E. (1993a) «Ai margini dell'Europa: problemi regionali delle due Irlande», *Atti del Convegno "1992 e Periferie d'Europa"* (Lecce, 14-16 gennaio 1993): 63-86

BIAGINI E. (1993b) «Spatial dimensions of conflict», *GeoJournal*, 31, 2: 119-128

BIRD J. (1977) *Centrality and cities*, London, Routledge & Kegan Paul

BIRD J. (1980) «Seaports as a subset of gateways for regions: a research survey», *Progress in Human Geography*, IV, 3: 360-370

BIRD J. (1981) «Gateways: slow recognition but irresistible rise», *Tijdschrift voor Economische en Sociale Geografie*, LXXIV, 3: 196-202

BLACKER C.P. (1947) «Stages in population growth», *Eugenics Review*, 88-102

BOUDON R. (1984) *La place du désordre. Critique des théories du changement social*, Paris, PUF

BOULDING K.E. (1956) «General systems theory: the skeleton of science», *General Systems*, 1: 11-17

BRAUDEL F. (1965) *Civiltà e imperi del Mediterraneo nell'età di Filippo II*, 2 vols., Torino, Einaudi, 2nd edn. (transl. from French into Italian)

BRAUDEL F. (1977) *Civiltà materiale, economia e capitalismo (secoli XV-XVIII)*, 3 vols., Torino, Einaudi (transl. from French into Italian)

CHRISTOPHER A.J. (1988) *The British Empire at its zenith*, London, Croom Helm

DARWIN C. (s.d.) *The Origin of Species and The Descent of Man*, New York, The Modern Library (reprints of the originals published respectively in 1859 and 1871)

DJILAS M. (1957) *The new class: an analysis of the communist system*, New York, Praeger

FRIEDMANN J. (1966) *Regional development policy: a case study of Venezuela*, Cambridge, Mass., MIT Press

FRIEDMANN J. (1972) «A general theory of polarized development», in *Growth centers in regional economic development*, ed. N.M. Hansen, New York, Free Press: 82-107

GLUCKSMANN A. & WOLTON T. (1987) *Silenzio, si uccide*, Milano, Longanesi (transl. from French into Italian)

HUIZINGA J. (1957) *Herfsttij der Middeleeuwen*, 9th ed., Haarlem, Tjeenk

HUGGETT R. (1980) *Systems analysis in geography*, Oxford, Clarendon

KENWOOD A.G. & LOUGHEED A.L. (1983) *The growth of the international economy 1820-1980*, London, Allen & Unwin

LENIN V.I. (1916) *Imperializm, kak vyssaja stadija kapitalizma*, Petrograd, Parus

LENIN V.I. (n.d.) *The revolutionary proletariate and the right of nations to self-determination*, Moscow, Progress Publishers (transl. from Russian into English)

LIVINGSTON M.H. (ed.) (1978) *International terrorism in the contemporary world*, London, Greenwood Press

MATHIEU V. (1973) *La speranza nella rivoluzione*, Milano, Rizzoli

MAY B. (1981) *The Third World calamity*, London, Routledge & Kegan Paul

McEVEDY C. & JONES A.L. (1983) *Atlas of world population history*, Harmondsworth, Penguin

NOTESTEIN F.W. (1953) «The economics of population and food supply», *Proceedings of the 8th International Conference of Agricultural Economists*, London, Oxford University Press

PARSONS T. (1951) *The social system*, New York, Free Press

PARSONS T. (1951) *The system of modern societies*, Englewood Cliffs, N.J., Prentice Hall

PELLICANI L. (1984) *Miseria del marxismo*, Milano, Sugarco

PERA M. (1979) «È scientifico il programma scientifico di Marx?», *Studium*, 4: 441-463

PERROUX F. (1955) «Note sur la notion de "pôle de croissance"», *Economie appliquée*, 8: 307-320

PINDER D. (1983) *Regional economic development and policy*, London, Allen & Unwin

POPPER K. (1963) *The open society and its enemies*, 2 vols., London, Routledge & Kegan Paul

POPPER K. (1968) *The logic of scientific discovery*, London, Oxford University Press

RAPOPORT A. et al. (1968) «Systems analysis», in *International Encyclopedia of the Social Sciences*, ed. D.L.Sills, New York, Macmillan & Free Press, vol. 15: 452-495

SAMEK LODOVICI E. (1991) *Metamorfosi della gnosi. Quadri della dissoluzione contemporanea*, Milano, Ares, 2nd edn.

STORR A. (1968) *Human aggression*, Harmondsworth, Penguin

TOCQUEVILLE A. DE (1992) *La democrazia in America*, Milano, BUR (transl. from French into Italian)

THOMPSON W.S. (1946) *Population and peace in the Pacific*, Chicago, University of Chicago Press

VON BERTALANFFY L. (1968) *General system theory*, New York, Braziller

WIGLEY T.M.L. et al. (eds.) (1981) *Climate and history: studies on past climates and the impact on man*, Cambridge, Cambridge University Press

WITTFOGEL K.A. (1957) *Oriental despotism: a comparative study of total power*, New Haven, Conn., Yale University Press

All Ireland: geography

AALEN F.H.A. (1978) *Man and the landscape in Ireland*, London - New York - San Francisco, Academic Press

AALEN F.H.A. (1986) «The rehousing of rural labourers in Ireland under the Labourers (Ireland) Acts, 1883-1919», *Journal of Historical Geography*, 12, 3: 287-306

AALEN F.H.A. (1988) «Homes for Irish heroes: housing under the Irish Land (Provision for Soldiers and Sailors) Act 1919, and the Irish Sailors' and Soldiers' Trust», *Town Planning Review*, 59, 3: 305-323

AALEN F.H.A. (1989) «Imprint of the past», in *The Irish countryside*, ed. D. Gillmor, Dublin, Wolfhound Press: 83-120

ALMAGIÀ R. (1932) «La cartografia e la geografia descrittiva dell'Irlanda nel secolo XVII», *Rivista Geografica Italiana*, XXXIX, IV-V: 138-141

ANDREWS K.R. et al. (1979) *The Westward enterprise: English activities in Ireland, the Atlantic and America, 1480-1650*, Detroit, Wayne State University Press

Antipode (1980) 11, 2, *Special issue on Ireland* (contributions by C.Regan, J.MacLaughlin, D.G.Pringle, F.W.Boal, J.Anderson, D.Perrons, F.Walsh, D.Parsons, D.Byrne, P.J.Duffy, S.Daultrey, P.M.Goodyear)

BANNON M.J. (undated) «Urban and regional planning in the Republic of Ireland», in «Planning in Ireland: an overview», *Occasional Papers in Planning*, No. 1, Department of Town and Country Planning, Belfast, The Queen's University: 1-20

BARRETT G. (1982) «Problems of spatial and temporal continuity of rural settlement in Ireland, A.D. 400-1169», *Journal of Historical Geography*, 8, 3: 245-260

BENDER R.J. (ed.) (1984) *Neuere Forschungen zur Sozialgeographie von Irland*, Mannheim, Mannheimer Geographische Arbeiten

BIAGINI E. (1992) *Irlanda. Sviluppo e conflitto alla periferia d'Europa*, Genova, Nuove Edizioni del Giglio

BIAGINI E. (1993) «Ai margini dell'Europa: problemi regionali delle due Irlande», *Atti del Seminario Internazionale 1992 e periferie d'Europa*, eds. C. Santoro Lezzi & A. Trono, Bologna, Pàtron: pp. 199-222

BONASERA, F. (1966) «L'Irlanda e la sua economia», *L'Universo*, 46, 2: 761-796

BRUNT B. (1988) *The Republic of Ireland*, London, Chapman

BRUNT B. (1989) «The new industrialisation of Ireland», in *Ireland: contemporary perspectives on a land and its people*, eds. R.W.G. Carter & A.J. Parker, London - New York, Routledge: 201-236

BOSWELL J. (1953) *Life of Johnson*, Oxford, Oxford University Press (first published in 1791)

BUCHANAN C. & PARTNERS (1968) *Regional studies in Ireland*, Commissioned by the United Nations on behalf of the Government of Ireland, September

BUTLIN R.A. (ed.) (1977) *The development of the Irish town*, London, Croom Helm - Totowa, N.J., Rowman & Littlefield

CARTER B. (1989) «Resources and management of Irish coastal waters and adjacent coasts», in *Ireland: contemporary perspectives on a land and its people*, eds. R.W.G. Carter & A.J. Parker, London - New York, Routledge: 393-419

CARTER R.W.G. & PARKER A.J. (eds.) (1989) *Ireland: contemporary perspectives on a land and its people*, London - New York, Routledge

CAWLEY M. (1989a) «Rural people and services», in *The Irish countryside*, ed. D. Gillmor, Dublin, Wolfhound Press: 197-225

CAWLEY M. (1989b) «The problems of rural Ireland», in *Ireland: contemporary perspectives on a land and its people*, eds. R.W.G. Carter & A.J. Parker, London - New York, Routledge: 145-170

CAWLEY M.E. (1990) «Population change in the Republic of Ireland», *Area*, 22, 1: 67-74

COGHLAN R. (1985) *Pocket dictionary of Irish myth and legend*, Belfast, Appletree Press

CONNELL K.H. (1950) *The population of Ireland 1750-1845*, Oxford, Clarendon

CONVERY F.G. (1989) «European economic policies and Ireland», in *Ireland: contemporary perspectives on a land and its people*, eds. R.W.G. Carter & A.J. Parker, London - New York, Routledge: 9-21

CORBEL J. (1952) «Une région karstique d'Irlande: le Burren», *Revue de Géographie de Lyon*, 27: 21-33

COWARD J. (1982) «Fertility changes in the Republic of Ireland during the 1970s», *Area*, 14, 2: 109-117

COWARD J. (1989) «Irish population problems», in *Ireland: contemporary perspectives on a land and its people*, eds. R.W.G. Carter & A.J. Parker, London - New York, Routledge: 55-86

DREW D. (1989) «The shape of the land», in *The Irish countryside*, ed. D. Gillmor, Dublin, Wolfhound Press: 15-48

EDWARDS R. D. & WILLIAMS T.D. (eds.) (1956), *The Great Famine*, Dublin, Browne & Nolan

EVANS E.E. (1964) «Ireland and Atlantic Europe», *Geographische Zeitschrift*, 52: 224-241

EVANS E.E. (1973) *The personality of Ireland*, Cambridge, University Press

FLATRES A. (1964) «Une étude socio-démographique sur la campaigne ir-landaise», *Annales de Géographie*, 73: 595-598

FREEMAN T.W. (1957) *Pre-Famine Ireland: a study in historical geography*, Manchester, Manchester University Press

FREEMAN T.W. (1960) *Ireland: a general and regional geography*, London, Methuen, 2nd edn.

GILL C. (1964) *The rise of the Irish linen industry*, Oxford, Oxford University Press, reprint

GILLMOR D.A. (ed.) (1979) *Irish resources and land use*, Dublin, Institute of Public Administration

GILLMOR D.A. (1989) «Agricultural development», in *Ireland: contemporary perspectives on a land and its people*, eds. R.W.G. Carter & A.J. Parker, London - New York, Routledge: 171-199

GILLMOR D.A. (ed.) (1989) *The Irish countryside*, Dublin, Wolfhound Press

GUIFFAN J.-J.V. (1970) *L'Irlande: milieu et histoire*, Paris, Colin

HUTTERMANN A. (1978) *Industrieparks in Irland*, Wiesbaden, Steiner

JOHNSON J.H. (1990) «The context of migration: the example of Ireland in the nineteenth century», *Transactions of the Institute of British Geographers*, New Series, 15, 3: 259-276

JOYCE P.W. (1984) *Pocket guide to Irish place names*, Belfast, Appletree Press

KILLEN J. & SMYTH A. (1989) «Transportation», in *Ireland: contemporary perspectives on a land and its people*, eds. R.W.G. Carter & A.J. Parker, London - New York, Routledge: 271-300

LLOYD PRAEGER R. (1953) *Irish landscape*, Cork, Cultural Relations Committee of Ireland

McCABE A.M. (1993) «The 1992 Farrington Lecture: Drumlin Bedforms and Related Ice-Marginal Depositional Systems in Ireland», *Irish Geography*, 26: 22-44

McCRACKEN E. (1971) *The Irish woods since Tudor times: their distribution and exploitation*, Newton Abbot, David & Charles

MITCHELL F. (1976) *The Irish landscape*, London, Collins

NEWBOULD P.J. (1989) «Irish energy: problems and prospects», in *Ireland: contemporary perspectives on a land and its people*, eds. R.W.G. Carter & A.J. Parker, London - New York, Routledge: 331-358

NEWMANN J. (1967) *New dimensions in regional planning: a case study of Ireland*, Dublin, An Forhas Forbartha

NOLAN W. (ed.) (1986) *The shaping of Ireland: the geographical perspective*, Cork & Dublin, Mercier Press

O'FARRELL P.N. (1975) *Regional industrial development trends in Ireland 1960-1973*, Dublin, Industrial Development Authority (IDA), Ireland, Paper 1

O'FARRELL P.N. (1978) «An analysis of industrial location: the Irish case», *Progress in Planning*, 9: 129-229

ORME A.R. (1970) *Ireland*, «The World's Landscapes», 4, London, Longman

PARKER T. (1989) «The changing nature of Irish retailing», in *Ireland: contemporary perspectives on a land and its people*, eds. R.W.G. Carter & A.J. Parker, London - New York, Routledge: 237-270

POLLARD J. (1989) «Patterns in Irish tourism», in *Ireland: contemporary perspectives on a land and its people*, eds. R.W.G. Carter & A.J. Parker, London - New York, Routledge: 301-330

PRINGLE D. (1989) «Partition, politics and social conflict», in *Ireland: contemporary perspectives on a land and its people*, eds. R.W.G. Carter & A.J. Parker, London - New York, Routledge: 23-54

ROHAN P.K. (1975) *The climate of Ireland*, Dublin, Stationery Office

ROTTMAN D. (1989) «Crime in geographical perspective», in *Ireland: contemporary perspectives on a land and its people*, eds. R.W.G. Carter & A.J. Parker, London - New York, Routledge: 87-111

ROYLE S.A. (1989) «The historical legacy in modern Ireland», in *Ireland: contemporary perspectives on a land and its people*, eds. R.W.G. Carter & A.J. Parker, London - New York, Routledge: 113-143

SIMMS A. (1979) «Irland: Überformung eines keltischen Siedlungsraumes am Rande Europas durch externe Kolonisationsbewegungen», in *Gefügemuster der Erdoberfläche: Die genetische Analyse von Reliefkomplexen und Siedlungsräumen*, eds. J. Hagedorn et al., Festschrift zum 42. Deutschen Geographentag, Göttingen, Verlag Erich Goltze: 261-308

SMYTH W.J. (1978) «The western isle of Ireland and the eastern seaboard of America: England's frontiers», *Irish Geography*, 11: 1-22

SMYTH W.J. (1978) «Landholding changes, kinship networks and class transformation in rural Ireland: a case study from County Tipperary», *Irish Geography*, 16: 16-35

STEPHENS N. & GLASSCOCK R.E. (eds.) (1970) *Irish geographical studies in honour of E. Estyn Evans*, Belfast, Queen's University, Department of Geography

SWEENEY J. (1985) «The changing synoptic origins of Irish precipitation», *Transactions of the Institute of British Geographers*, New Series, 10, 4: 467-480

SWEENEY J. (1989) «Air pollution problems in Ireland», in *Ireland: contemporary perspectives on a land and its people*, eds. R.W.G. Carter & A.J. Parker, London - New York, Routledge: 421-439

VERRIÈRE J. (1978) *La population de l'Irlande*, Paris, Mouton

WALSH J.A. (1989) «Regional development strategies», in *Ireland: contemporary perspectives on a land and its people*, eds. R.W.G. Carter & A.J. Parker, London - New York, Routledge: 441-471

WHELAN K. (1988) «The impact of Irish Catholicsm 1700-1850», in *Common ground: essays on the historical geography of Ireland*, eds. W. Smyth & K. Whelan, Cork, Cork University Press: 253-277

All Ireland: history, society, politics, economy

AA.VV. (1986) *Ireland after the Union*, Proceedings of the 2nd joint meeting of the Royal Irish Academy and the British Academy, Oxford, University Press

ADAIR R.A. (1886) *Ireland and her servile war*, London, Ridgway

ALISON W.P. (1847) *Observations on the famine of 1846-47*, Edinburgh, Blackwood

ALMQUIST E.L. (1979) «Pre-famine Ireland and the theory of European proto-industrialization: evidence from the 1841 census», *Journal of Economic History*, 3: 699-718

BAGWELL R. (1963) *Ireland under the Tudors*, 3 vols., London, Holland

BAILEY I.R. & SHEEHY S.J. (1971) *Irish agriculture in a changing world*, Edinburgh, Oliver & Boyd

BARNARD T.C. (1975) *Cromwellian Ireland: English government and Reform in Ireland 1649-1660*, Oxford, Oxford University Press

BARTELETT T. & HAYTON D.W. (eds.) (1979) *Penal era and golden age*, Belfast, Ulster Historical Foundation

BECKETT J.C. (1966) *The making of modern Ireland*, London, Faber & Faber

BECKETT J.C. (1983) *The Anglo-Irish tradition*, Belfast, Blackstaff Press, reprint

BENNET R. (1961) *The Black and Tans*, London, Four Square Books, reprint

BEW P. (1987) *Conflict and conciliation in Ireland 1890-1910*, Oxford, Clarendon

BICHENO J.E. (1830) *Ireland and its economy*, London, Murray

BIRNIE J.E. (1957) *An economic history of the British Isles*, London, Methuen

BOLTON J.C. (1966) *The passing of the Irish Act of Union*, Oxford, Oxford University Press

BOTTIGHEIMER K.S. (1971) *English money and Irish land: the "Adventurers" in the Cromwellian settlement*, Oxford, Clarendon

BOWEN D. (1978) *The Protestant crusade in Ireland 1800-70*, Dublin, Gill & Macmillan

BOWEN K. (1983) *Protestants in a Catholic state: Ireland's privileged minority*, Montreal, McGill-Queen's University Press

BOWYER BELL J. (1972) *The secret army: a history of the IRA 1915-1970*, London, Sphere Books, reprint

BOYCE D.G. (1982) *Nationalism in Ireland*, London, Croom Helm

BOYCE D.G. (1988a) *The Irish question and British politics 1868-1986*, Basingstoke, Macmillan

BOYCE D.G. (1988b) *The Revolution in Ireland*, Dublin, Gill & Macmillan

BOYD A. (1972) *The rise of the Irish Trade Unions 1729-1970*, Tralee, Co. Kerry, Anvil Books

BOYLE R. (1882-92) *History of the Irish Confederation and the war in Ireland 1641-49*, 2 vols., London, Gilbert

BRADLEY D. (1988) *Farm labourers, Irish struggle 1900-1976*, Belfast, Athol Books

BROWN T. (1987) *Ireland: a social and cultural history*, London, Fontana, 4th reprint

BUCKLAND P. (1972) *Irish Unionism: the Anglo-Irish and the New Ireland*, Dublin, Gill & Macmillan; New York, Barnes & Noble Books

BURKE J.F. (undated) *Outlines of the industrial history of Ireland*, Dublin, Browne & Nolan

CANNY N. (1991) «Early modern Ireland, c. 1500-1700», in *The Oxford illustrated history of Ireland*, ed. R.F. Foster, Oxford - New York, Oxford University Press, reprint: 104-160

CARTER R.W. et al. (1993) «Man's impact on the coast of Ireland», in *Tourism vs Environment: the case for coastal areas*, ed. Wong P.P., *The GeoJournal Library*, Dordrecht-Boston-London, Kluwer Academic Publishers: 211-225

CAULFIELD M. (1965) *The Easter rebellion*, London, Four Square Books

CHAUVIRÉ R. (1949) *Histoire d'Irlande*, Paris, PUF

CHURCHILL W. (1962) *A history of the English-speaking peoples*, 4 vols., London, Cassell, reprint

CLANCY P. et al. (eds.) (1986) *Ireland: a sociological profile*, Dublin, Institute of Public Administration, Sociological Society of Ireland

CLARKE A. (1968) «The Graces 1625-41», *Irish History Series*, 8, Dundalk, Dundalgan Press

COLLISON BLACK R.D. (1960) *Economic thought and the Irish question 1817-1870*, Cambridge, University Press

CONNELL K.H. (1962) «Peasant marriage in Ireland: its structure and development since the Famine», *Economic History Review*, 2nd series, 3: 502-523

COOGAN T.P. (1970) *The IRA*, London, Pall Mall Press

CULLEN L.M. (1972) *An economic history of Ireland since 1660*, London, Batsford

CURRY J. (1786) *An historical and critical review of the civil wars in Ireland from the reign of Queen Elizabeth to the settlement under King William, with the state of the Irish Catholics*, 2 vols., Dublin, White

CURTIS E. (1936) *A history of Ireland*, London, Methuen

DAVIES R. R. (1990) *Domination and conquest: the experience of Ireland, Scotland and Wales*, Cambridge, Cambridge University Press

DAVITT M. (1904) *The fall of feudalism or the story of the Land League revolution*, London, Harper

DELANEY T.G. (1977) «The archaeology of the Irish town», in *European towns: their archaeology and early history*, ed. M.W. Barley, London - New York - San Francisco, Academic Press: 47-64

DE PAOR L. (1986) *The peoples of Ireland: from prehistory to modern times*, London, Hutchinson

DE PAOR L. (1990) *Unfinished business: Ireland today and tomorrow*, London, Radius

DE PAOR M. & L. (1958) *Early Christian Ireland*, London, Thames & Hudson

DE VERE WHITE T. (1966) *Kevin Higgins: strong man in the Free State government: 1922-27. His eventful life and mysterious violent death*, Tralee, Co.Kerry, Anvil Books, reprint

DRAKE M. (1963) «Marriage and population growth in Ireland 1750-1845», *Economic History Review*, 2nd series, 2: 301-313

DUNLOP R. (1913) *Ireland under the Commonwealth*, 2 vols., Manchester, Manchester University Press

DUFF C. (1970) *La rivolta irlandese (1916-1921)*, Milano, Rizzoli (transl. from English into Italian)

EDGEWORTH M. (1964) *Castle Rackrent*, Oxford, Oxford University Press (first published in 1800)

EDWARDS R. D. (1977a) *Ireland in the age of the Tudors: the destruction of Hiberno-Roman civilization*, London, Croom Helm

EDWARDS R. D. (1977b) *Patrick Pearse: the triumph of failure*, London, Faber & Faber

ELLIOTT M. (1978) «The origins and transformation of early Irish republicanism», *International Review of Social History*, 3: 405-428

ELLIOTT M. (1989) *Wolfe Tone, prophet of Irish independence*, New Haven, Conn. - London, Yale University Press

ELLIS P.B. (1975) *Hell or Connaught: the Cromwellian colonisation of Ireland 1652-1660*, London, Hamish Hamilton

ELLY S. (1848) *Potatoes, pigs and politics*, London, Kent & Richards

FANNING R. (1989) «Britain, Ireland and the end of Union», in *Ireland after the Union: Proceedings of the Second Joint Meeting of the Royal Irish Academy and the British Academy* (London, 1986), Oxford, Oxford University Press: 105-120

FENNEL D. (1989) *The revision of Irish nationalism*, Dublin, Open Air

FITZPATRICK D. (1991) «Ireland since 1870», in *The Oxford illustrated history of Ireland*, ed. R.F. Foster, Oxford - New York, Oxford University Press, reprint: 213-274

FISHER J. (1862) *How Ireland may be saved*, London, Ridgway

FLANAGAN M.T. (1989) *Irish society, Anglo-Norman settlers, Angevin kingship*, Oxford, Clarendon

FOSTER R.F. (1991) «Ascendancy and Union», in *The Oxford illustrated history of Ireland*, ed. R.F. Foster, Oxford - New York, Oxford University Press, reprint: 161-212

FOSTER R.F. (ed.) (1991) *The Oxford illustrated history of Ireland*, Oxford - New York, Oxford University Press, reprint

FRAME R. (1982) *English lordship in Ireland 1318-1361*, Oxford, Clarendon

FRAME R. (1990) *The political development of the British Isles 1100-1400*, Oxford, University Press

FRASER A. (1973) *Cromwell, our chief of men*, London, Weidenfeld & Nicholson

FRASER T.G. (1973) *Partition in Ireland, India and Palestine: theory and practice*, London, Macmillan

FROUDE J.A. (1881) *The English in Ireland*, London, Longmans

GALIBERTI L. (1990) «Risorgimento italiano e movimento nazionalista irlandese», *Daniel O'Connell, Atti of the Convegno di studi nel 140° anniversario della morte*, ed. L. Morabito, Genova, Comune di Genova, Istituto Mazziniano: 81-92

GALLAGHER F. (1957) *The indivisible Ireland: the history of the partition of Ireland*, London, Gollancz

GLEESON J. (1963) *Bloody Sunday*, London, Four Square Books, reprint

GODKIN J. (1870) *Land war in Ireland*, London, Macmillan

GOLDSTROM J.M. & CLARKSON L.A. (eds.) (1981) *Irish population, economy and society*, Oxford, Clarendon

GRAHAM B.J. & PROUDFOOT L.J. (eds.) (1993) *An historical geography of Ireland*, London, Academic Press

GREEN A.S. (1909) *The making of Ireland*, London, Macmillan

GREEN R. et al. (1990) *Understanding contemporary Ireland. State, class and development in the Irish Republic*, Basingstoke, Gill & Macmillan

GUINNESS D. & DONOHOGUE D. (1986) *Ascendancy Ireland*, Los Angeles, Cal., UCLA, Clark Libray Seminar

GWYNN D. (1947) *Daniel O'Connell*, Cork, University Press

HACHEY T.E. & MAC CAFFREY L.J. (eds.) (1988) *Perspectives on Irish nationalism*, Lexington, Ky., University of Kentucky Press

HACHEY T.E. et al. (1989) *The Irish experience*, Englewood Cliffs, N.J., Prentice-Hall

HAGAN G. (1913) *Home Rule: l'autonomia irlandese*, Roma, Max Bretschneider (transl. from English into Italian)

HECHTER M. (1975) *Internal colonialism: the Celtic fringe in British national development 1536-1966*, London, Routledge & Kegan Paul

HERITY M. & EOGAN G. (1977) *Ireland in prehistory*, London - New York, Routledge

HICKSON M. (1884) *Ireland in the seventeenth century and massacre of 1641*, London, Longmans

HUTCHINSON J. (1987) *The dynamics of cultural nationalism*, London, Allen & Unwin

INGLIS T. (1987) *Moral monopoly: the Catholic Church in modern Irish society*, Dublin, Gill & Macmillan

JACKSON H. (1972) «The two Irelands», in *The Fourth World: victims of group oppression*, ed. B. Whitaker, London, Sidgwick & Jackson: 188-216

JOYCE P.W. (1903) *A social history of ancient Ireland*, London, Longmans

KEATING G. (1902-1914) *The history of Ireland*, 4 vols., London, Nutts

KEE R. (1972) *The green flag: a history of Irish nationalism*, 3 vols., London, Weidenfeld & Nicholson

KEE R. (1982) *Ireland: a history*, London, Abacus, reprint

KELLY J.S. (1987) *The Bodyke evictions*, Scariff, Co. Clare, Fossabeg Press

KENNEDY K.A. et al. (1988) *The economic development of Ireland in the twentieth century*, London - New York, Routledge

KENNEDY R.E. jr. (1973) *The Irish: emigration, marriage and fertility*, Berkeley, Cal., University of California Press

KENNEDY G. & GOOD M. (eds.) (1984) *The guide to political parties in Ireland*, Killala, Co. Mayo, Morrigan

KINEALY C. (1994) *This great calamity. The Irish Famine 1845-52*, Dublin, Gill & Macmillan

LARKIN E. (1965) *James Larkin, 1876-1947, Irish labour leader*, London, Routledge & Kegan Paul

LARKIN E. (1975) *The Roman Catholic Church and the creation of the modern Irish state 1878-1886*, Dublin, Gill & Macmillan

LARKIN E. (1976) *The historical dimensions of Irish Catholicism*, New York, Arno

LAWLOR S. (1983) *Britain and Ireland 1914-23*, Dublin - Totowa, N.J., Gill & Macmillan

LECKY W.E.H. (1892) *A history of Ireland in the eighteenth century*, 5 vols., London, Longmans

LEE J. et al. (1966) «Money and beer in Ireland 1790-1875», *Economic History Review*, 2nd series, 1: 183-194

LEE J.J. (1990) *Ireland 1912-1985; politics and society*, Cambridge, University Press

LENIN V.I. (1970) *Lenin on Ireland*, Dublin & Belfast, New Books Publications

LONGFIELD A.K. (1929) *Anglo-Irish trade in the sixteenth century*, London, Routledge

LONGFORD, EARL OF & O'NEILL T.P. (1970) *Eamon De Valera*, Dublin, Gill & Macmillan

LOUGH E. (1896) *England's wealth and Ireland's poverty*, London, Fisher & Unwin

LUING S,O. (1970) *I die in a good cause*, Tralee, Co.Kerry, Anvil Books

LYDON J. (ed.) (1981) *England and Ireland in the later Middle Ages*, Kill Land, Blackrock, Co. Dublin, Irish Academic Press

LYDON J. & MAC CURTAIN M. (eds.) (1972-75) *The Gill history of Ireland*, 11 vols., Dublin, Gill & Macmillan

LYONS F.S.L. (1973) *Ireland since the Famine*, Glasgow, Collins

LYONS F.S. & HAWKINS R.A.J. (eds.) (1980) *Ireland under the Union*, Oxford, Clarendon

MacCAFFREY L.J. (1968) *The Irish question*, Lexington, Ky., University of Kentucky Press

MacCAFFREY L.J. (1979) *Ireland, from colony to nation-state*, Englewood Cliffs, N.J., Prentice-Hall

MacCRACKEN J.L. (1971) *The Irish Parliament in the eighteenth century*, «Irish History Series», No. 9, Dundalk, Dundalgan Press

MacDERMOT F. (1969) *Theobald Wolfe Tone and his times*, Tralee, Co.Kerry, Anvil Books, reprint

MacDOWELL R.B. (1957) *Social life in Ireland 1800-1845*, Cork, Cultural Relations Committee of Ireland

MacDOWELL R.B. (1979) *Ireland in the age of Imperialism and revolution 1760-1801*, Oxford, Clarendon

MacINTYRE A. (1965) *The Liberator: Daniel O'Connell and the Irish Party 1830-1847*, London, Hamish Hamilton

MacKENNA E.E. (1978) «Age, region and marriage in post-Famine Ireland: an empirical examination», *Economic History Review*, 2nd series, 2: 238-256

MacLYSAGHT E. (1969) *Irish life in the seventeenth century*, Shannon, Irish University Press - Cork, Cork University Press, reprint

MacMANUS M.J. (1939) *Irish cavalcade 1550-1850*, London, Macmillan

MANSERGH N. (1942) *Britain and Ireland*, London, Longmans

MANSERGH N. (1975) *The Irish question 1840-1921*, London, Allen & Unwin, 3rd ed.

MARLOW J. (1973) *Captain Boycott and the Irish*, London, Deutsch

MARX K. & ENGELS F. (1971) *Ireland and the Irish question*, Moscow, Progress Publishers

MAXWELL C. (1940) *Country and town in Ireland under the Georges*, London, Harrap

MEENAN J. (1970) *The Irish economy since 1922*, Liverpool, Liverpool University Press

MILLER D.W. (1973) *Church, State and Nation in Ireland 1898-1921*, Dublin, Gill & Macmillan

MITCHEL J. (1991) *Giornale di prigionia 1848-1853*, Bergamo, Lubrina, 2 vols. (transl. from English into Italian)

MOKYR J. (1980a) «Malthusian models and Irish history», *Journal of Economic History*, 4: 159-168

MOKYR J. (1980b) «The deadly fungus: an econometric investigation into the short-term demographic impact of the Irish Famine, 1846-1851», *Research in Population Economics*, 2: 237-277

MOKYR J. (1983) *Why Ireland starved: a quantitative and analytical history of the Irish economy 1800-1850*, London, Allen & Unwin

MOKYR J. & O' GRADA C. (1988) «Poor and getting poorer? Living standards in Ireland before the Famine», *Economic History Review*, 2nd series, 2: 209-235

MOODY T.W. (1976) *A new history of Ireland*, Oxford, Clarendon

MOODY T.W. & MARTIN F.X. (1967) *The course of Irish history*, Cork, Mercier Press

MOODY T.W. & VAUGHAN W.E. (1986) *A new history of Ireland: eighteenth century Ireland 1691-1800*, Oxford, Clarendon

MORABITO L. (ed.) (1990) *Daniel O'Connell, Atti of the Convegno di studi nel 140° anniversario della morte*, Genova, Comune di Genova, Istituto Mazziniano

MORGAN A. & PURDIE B. (eds.) (1980) *Ireland: divided nation, divided class*, London, Ink Links

NICHOLAS S.G. & DZIEGIELEWSKI M. (1980) «Supply elasticities, rationality and structural change in Irish agriculture 1850-1925», *Economic History Review*, 2nd series, 3: 411-414

NORMAN E.R. (1965) *The Catholic Church and Ireland in the age of rebellion 1859-1873*, London, Longmans

NORMAN E.R. (1969) *The Catholic Church and Irish politics in the eighteenth century*, «Irish History Series», No. 5, Dundalk, Dundalgan Press

O'BEIRNE RANELAGH J. (1983) *A short history of Ireland*, Cambridge, Cambridge University Press

O'BRIEN G. (1918) *Economic history of Ireland in the eighteenth century*, Dublin, Maunsel

O'BRIEN G. (1919) *Economic history of Ireland in the seventeenth century*, Dublin, Maunsel

O'BRIEN G. (1921) *Economic history of Ireland from the Union to the Famine*, London, Longmans

O'BROIN L. (1958) *The unfortunate life of Mr. Robert Emmet*, Dublin, Clonmore & Reynolds

O'CONNOR F. (1969) *The big fellow: Michael Collins and the Irish revolution*, London, Gorgi Books, reprint

O'CORRÀIN D. (1991) «Prehistoric and early Christian Ireland», in *The Oxford illustrated history of Ireland*, ed. R.F. Foster, Oxford - New York, Oxford University Press, reprint: 1-52

O'DONOHOGUE F. (1954) *No other law: the story of Liam Lynch and the IRA 1916-23*, Dublin, Irish Press

O'DONOHOGUE F. (1958) *Tomás MacCurtain*, Tralee, Co.Kerry, Kerryman

O'DONOVAN ROSSA J. (1967) *My years in English jails: the brutal facts*, ed. Sean Ua Cearnaigh, Tralee, Co.Kerry, Anvil Books, reprint

O'GRADA C. (1975) «Supply responsiveness in Irish agriculture during the nineteenth century», *Economic History Review*, 2nd series, 2: 312-317

O'GRADA C. (1980) «Supply elasticities in Irish agriculture: a reply», *Economic History Review*, 2nd series, 3: 415-416

O'GRADA C. (1988) *Ireland before and after the Famine*, Manchester, Manchester University Press

O'GRADA C. (1989) *The Great Irish Famine*, Basingstoke, Macmillan

O'HAGAN J.W. (ed.) (1981) *The economy of Ireland: policy and performance*, Dublin, Irish Management Institute, 3rd edn.

O'HALPIN E. (1987) *The decline of the Union: British government in Ireland 1892-1920*, Dublin, Gill & Macmillan

O'HEGARTY P.S. (1952) *A history of Ireland under the Union 1801-1922*, London, Methuen

O'LEARY C. et al. (1988) *The Northern Ireland Assembly 1982-1986*, London, Hurst; Belfast, Queen's University Bookshop

O'LEARY J. (1968) *Recollections of Fenians and Fenianism*, 2 vols., Shannon, Irish University Press

O'MAHONY D. (1970) *The Irish economy*, Cork, Cork University Press, reprint of the 2nd edn.

O'ROURKE K. (1991) «Did the Great Irish Famine matter?», *Journal of Economic History*, 51, 1: 1-22

O'ROURKE J. (1875) *History of the Great Irish Famine*, Dublin, McGlashan & Gill

OTWAY-RUTHVEN A.J. (1980) *A history of medieval Ireland*, London, Benn, 2nd edn.

PAKENHAM F. (1935) *Peace by ordeal: an account from first hand sources of the negotiation and signature of the Anglo-Irish Treaty 1921*, London, Jonathan Cape

PATTERSON H. (1989) *The politics of illusion: republicanism and socialism in modern Ireland*, London, Hutchinson Radius

PLOWDEN F. (1803) *An historical review of the state of Ireland from Henry II to the Union*, 2 vols., London

PRENDERGAST J.P. (1922) *The Cromwellian settlement of Ireland*, Dublin, Mellifont

RICHARDSON H.G. & SAYLES G.O. (1964) «Parliament in medieval Ireland», *Irish History Series*, 1, Dundalk, Dundalgan Press

ROBINSON O. (1962) «The London companies as progressive landlords in nineteenth century Ireland», *Economic History Review*, 2nd series, 1: 103-118

SCHLESINGER J.A. (1978) *Putting "Reality" together*, London, Constable

SILKE J.J. (1966) «Ireland and Europe 1559-1607», *Irish History Series*, 7, Dundalk, Dundalgan Press

SIMINGTON R.C. (1970) *The transplantation to Connacht 1654-58*, Shannon, Irish University Press

SIMMS J.G (1956) *The Willamite confiscation of Ireland 1690-1703*, London, Faber & Faber

SIMMS J.G. (1965) «The treaty of Limerick», *Irish History Series*, 6, Dundalk, Dundalgan Press

SIMMS J.G. (1966) «The Jacobite Parliament of 1689», *Irish History Series*, 2, Dundalk, Dundalgan Press

SIMMS J.G. (1969) *Jacobite Ireland 1685-91*, London, Routledge & Kegan Paul; Toronto, University of Toronto Press

SIMMS K. (1991) «The Norman invasion and Gaelic recovery», in *The Oxford illustrated history of Ireland*, ed. R.F. Foster, Oxford - New York, Oxford University Press, reprint: 53-103

STEPHENS J. (1965) *The insurrection in Dublin*, Dublin, Maunsel, reprint

TIERNEY M. (1976) *Croke of Cashel: the life of Archbishop Thomas Croke 1823-1902*, Dublin, Gill & Macmillan

TONE W.T.W. (1826) *Life of Theobald Wolfe Tone*, 2 vols., Washington, D.C.

TUCKER G.S.L. (1970) «Irish fertility ratios before the Famine», *Economic History Review*, 2: 267-284

WALSH B. (1970) «Religion and demographic behaviour in Ireland», Dublin, *Papers of the Economic and Social Research Institute*, No. 55

WHITE J. (1975) *Minority report: the protestant community in the Irish Republic*, Dublin, Gill & Macmillan

WHYTE J.H. (1971) *Church and State in modern Ireland*, Dublin, Gill & Macmillan

WHYTE J.H. (1972) «The Tenant League and Irish politics in the eighteen-fifties», *Irish History Series*, No. 4, Dundalk, Dundalgan Press

WINSTANLEY M.J. (1984) *Ireland and the land question 1800-1922*, Lancaster Pamphlets, London, Methuen

WOODHAM-SMITH C. (1962) *The Great Hunger: Ireland 1845-9*, London, Hamish Hamilton

YOUNGER C. (1988) *Ireland's civil war*, London, Fontana, 6th reprint

Travels and folklore

BUSH J. (1764) *Hibernia curiosa: giving a general view of the manners, customs, dispositions and co. of the inhabitants of Ireland*, London

CARR J. (1806) *The stranger in Ireland* (reprinted by Irish University Press, Shannon, 1970)

COGHLAN R. (1985) *Pocket dictionary of Irish myth and legend*, Belfast, Appletree Press

CROKER R.J.W. (1808) *A sketch of the state of Ireland, past and present*, Belfast, Appletree Press

DE BEAUMONT G.D. (1839) *L'Irlande sociale, politique et religieuse*, 2 vols., Paris

DE LACTONAYE ? (1796) *A Frenchman's walk through Ireland*, London (transl. from the French into English by J. Stevenson, 1917)

DUFFERIN LORD & BOYLE HON. G.F. (1847) *Narrative of a journey from Oxford to Skibbereen*, Oxford, Parker

FORBES J. (1852) *Memorandum made in Ireland*, London, Smith Elder

NICHOLSON A. (1844) *Ireland's welcome to the stranger*, London, Gilpin

O'FAOLAIN S. (1940) *An Irish journey*, London, Longmans

PELLEW G. (1852) *Memorandum made in Ireland*, London, Smith Elder

PETTY SIR W. (1671) *The political anatomy of Ireland* (reprinted by Irish University Press, Shannon, 1970)

POULETT SCROPE G. (1849) *Some notes of a tour in England, Scotland and Ireland*, London, Ridgway

REES A. & B. (1961) *Celtic heritage: ancient tradition in Ireland and Wales*, London, Thames & Hudson

REID T. (1823) *Travels in Ireland*, London, Longmans-Hunt

THOYER H. (ed.) (1699) *A trip to Ireland* (attributed to Edward Ward) Facsimile Texts Society, New York, Columbia University Press

TUKE J.H. (1880) *Journal. Irish distress and its remedies. A visit to Donegal and Connaught*, London, Ridgway

YEATS W.B. (1973) *Fairy and folk tales*, Dublin, Smythe

YOUNG A. (1780) *A tour in Ireland 1776-1779*, 2 vols. (reprinted by Irish University Press, Shannon, 1970)

The Irish overseas

AA.VV. (1980) *America and Ireland 1776-1976: the American identity and the Irish connection*, Proceedings of the United States Bicentennial Conference of Cumann, West, Conn. - London, Greenwood Press

BUSTEED M. (1996) «Dilemmas of identity in a displaced population. The Irish in Mid-Nineteenth Century Manchester» (unpublished paper at the RGS-IBG Conference, Glasgow, University of Strathclyde, 3rd-6th January)

DICKSON R.J. (1966) *Ulster emigration to colonial America*, London, Routledge & Kegan Paul

DUNCAN K. (1965) «Irish famine immigration and the social structure of Canada West», *Canadian Review of Sociology and Anthropology*, 2: 19-41

HANSEN M. (1940) *The Atlantic migration*, Cambridge, Mass., Harvard University Press

HAYES P. (1949) *Biographical dictionary of Irishmen in France*, Dublin, Gill

HENNESSY M.N. (1973) *The wild geese: the Irish soldier in exile*, London, Sidgwick & Jackson

HOUSTON C. & SMYTH W.J. (1978) «The Orange Order and the expansion of the frontier in Ontario 1830-1900», *Journal of Historical Geography*, 4: 251-264

JACKSON J.A. (1963) *The Irish in Britain*, London, Routledge & Kegan Paul

KIERNAN T.J. (1954) *The Irish exiles in Australia*, Dublin, Clonmore & Reynolds

McCAFFREY L.J. (1976) *The Irish diaspora in America*, Bloomington, Indiana University Press

MILLER K.A. (1985) *Emigrants and exiles: Ireland and the Irish exodus to North America*, New York, Oxford University Press

MOORE J. (1988) «Historical revisionism and the Irish in Britain», *Linen Hall Review*, 5, 3: 14-15

O'BRIEN G. (1940) *New light on emigration*, Cambridge, Mass., Harvard University Press

O'CONNOR K. (1972) *The Irish in Britain*, Dublin, Gill & Macmillan

SHAVEN W.V. (1963) *The American Irish: a political and social portrait*, New York, Macmillan

WALSH M. (ed.) (1961-64) *Spanish Knights of Irish origin*, 3 vols., Dublin, Stationery Office

Northern Ireland (geography, history, society, politics, economy)

ABBOTT D. (1973) «Ian Paisley: evangelism and confrontation in Northern Ireand», *Today's Speech*, 21: 49-55

ADAMS J. (1985) *Le finanze del terrorismo*, Milano, Sugarco (transl. from English into Italian)

ADAMS J. (1986) *The politics of Irish freedom*, Dingle, Brandon Books

ADAMS J. (1988) *A pathway to peace*, Cork, Mercier Press

ADAMSON I. (1974) *Cruthin: the ancient kindred*, Newtownards, Nosmada

ADAMSON I. (1981) *The identity of Ulster*, Belfast, The Author

AKENSON D.H. (1973) *Education and enmity. Control of schooling in Northern Ireland 1920-1950*, Newton Abbot, David & Charles

ARTHUR P. (1980) *Political realities: government and politics of Northern Ireland*, London, Longman

ARTHUR P. (1987) *Government and politics of Northern Ireland*, London, Longman, 2nd edn.

AUGHEY A. (1989) *Under siege: Ulster Unionism and the Anglo-Irish agreement*, Belfast, Blackstaff

AUNGER E.A. (1975) «Religion and occupational class in Northern Ireland», *Economic and Social Review*, 7: 1-18

BAMBERY C. (1986) *Ireland's permanent revolution*, London, Bookmarks

BARRITT D.P. & BOOTH A. (1972) *Orange and Green. A Quaker study of community relations in Northern Ireland*, Sedbergh, Yorks., Northern Friends Peace Board

BARRITT D.P. & CARTER C.F. (1962) *The Northern Ireland problem: a study in group relations*, London, Oxford University Press

BECKETT J.C. et al. (1972) *The Ulster debate*, London, Bodley Head

BELL G. (1976) *The Protestants of Ulster*, London, Pluto Press

BELL G. (1982) *Troublesome business: the Labour Party and the Irish question*, London, Pluto Press

BELL G. (1984) *The British in Ireland: a suitable case for withdrawal*, London, Pluto Press

BLACKMAN T. (1984) «Planning in Northern Ireland: the shape of things to come?», *The Planner*, 70, 1: 12-15

BLUMLER J.C. (1971) «Ulster on the small screen», *New Society*, 18: 1248-1250

BOAL F.W. (1974) *Social malaise in the Belfast Urban Area*, Belfast, Northern Ireland Community Relations Commission

BOAL F.W. & DOUGLAS J.N.H. (eds.) (1982) *Integration and division: geographical perspectives on the Northern Ireland problem*, London - New York, Academic Press

BOULTON D. (1973) *The UVF 1966-73: an anatomy of loyalist rebellion*, Dublin, Torc Books

BOYD A. (1972) *Brian Faulkner and the crisis of Ulster Unionism*, Tralee, Anvil Books

BOYLE K. & HADDEN T. (1994) *Northern Ireland: the choice*, Harmondsworth, Penguin

BRETT C.E.B. (1986) *Housing a divided community*, Dublin, Institute of Public Administration; Belfast, Institute of Irish studies, Queen's University

BROOKE P. (1987) *Ulster Presbyterianism*, Dublin, Gill & Macmillan

BROWN T. (1985) *The whole Protestant community: the making of a historical myth*, Derry, Field Day Pamphlet

BRUCE S. (1986) *God save Ulster! The religion and politics of Paisleyism*, Oxford, Clarendon

BUCHANAN R.H. (1982) «The planter and the Gael; cultural dimensions of the Northern Ireland problem», in *Integration and division: geographical perspectives on the Northern Ireland problem*, eds. F.W. Boal & J.N.H. Douglas, London - New York, Academic Press: 49-73

BUCKLAND P. (1981) *A history of Northern Ireland*, Dublin, Gill & Macmillan

BUCKLEY A.D. (1982) *A gentle people: a study of a peaceful community in Ulster*, Cultra, Co. Down, Ulster Folk and Transport Museum

BUCKLEY A.D. (1988) «Collecting Ulster's culture: are there really two traditions?», in *The use of tradition*, ed. A. Gailey, Cultra, Co. Down, Ulster Folk and Transport Museum: 49-60

BUFWACK M.F. (1982) *Village without violence: an examination of a Northern Irish community*, Cambridge, Mass., Schenkman

BUSTEED M.A. (1974) *Northern Ireland*, «Problem Regions of Europe», London, Oxford University Press

CAIRNS E. (1989) «Social identity and inter-group conflict in Northern Ireland: a developmental perspective», in *Growing up in Northern Ireland*, ed. J. Harbinson, Belfast, Stranmillis College, Learning Resources Unit: 115-130

CANNING D. et al. (1987) «Economic growth in Northern Ireland: problems and prospects» (unpublished paper submitted at the Economic and Social Research Council Conference on Economic and Social Research in Northern Ireland, Belfast, January)

CARTER R.W.G. (1993) «Age, origin and significance of the raised gravel barrier at Church Bay, Rathlin Island, County Antrim», *Irish Geography*, 26, 2: 141-146

COLLINS T. (1983) *The Centre cannot hold: Britain's failure in Northern Ireland*, Dublin, Bookworks Ireland

COMMON R. & GLASSCOCK R.E. (1964) «The regional geography of Ulster (Northern Ireland)», in *Field studies in the British Isles* (publ. for the 20th International Geographical Congress, United Kingdom), London, Nelson: 500-515

COMPTON P.A. (1976) «Religious affiliation and demographic variability in Northern Ireland», *Transactions of the Institute of British Geographers*, New Series, 1, 4: 433-452

COMPTON P.A. (1982a) «Fertility, nationality and religion in Northern Ireland», in *Demography of immigrants and minority groups in the United Kingdom*, ed. D.A. Coleman, London, Academic Press: 193-212

COMPTON P.A. (1982b) «The demographic dimension of integration and division in Northern Ireland», in *Integration and division: geographical perspectives on the Northern Ireland problem*, eds. F.W. Boal & J.N.H. Douglas, London - New York, Academic Press: 75-104

COMPTON P.A. (1985) «An evaluation of the changing religious composition of the population in Northern Ireland», *Economic and Social Review*, 16, 3: 201-224

COMPTON P.A. (1991a) «Demography: the 1980s in perspective» (unpublished paper)

COMPTON P.A. (1991b) «Employment differentials in Northern Ireland and job discrimination: a critique» (unpublished paper)

COMPTON P.A. & COWARD J. (1989) *Fertility and family planning in Northern Ireland*, Aldershot, Avebury

COMPTON P.A. & POWER J.F. (1986) «Estimates of the religious composition of Northern Ireland local government districts in 1981 and change in the geographical pattern of religious composition between 1971 and 1981», *Economic and Social Review*, 17, 2: 87-105

COMPTON P.A. et al. (1980) «Conflict and its impact on the urban environment of Northern Ireland», in *Development and settlement systems*, eds. G. Enyedi & J. Mészàros, Budapest, Akadémiai Kiadò: 83-98

CORMACK R.J. & OSBORNE R.D. (1989) «Employment and discrimination in Northern Ireland», *Policy Studies*, 9, 3: 49-54

CRUIKSHANK J.G. & WILCOCK D.N. (eds.) (1982) *Northern Ireland: environment and natural resources*, Belfast, The Queen's University

DALTON G. & MURRAY B. (1987) *Northern Ireland*, Cambridge, Cambridge University Press

DARBY J. (1976) *Conflict in Northern Ireland: the development of a polarised community*, Dublin, Gill & Macmillan

DARBY J. (ed.) (1983a) *A register of economic and social research on Northern Ireland*, London, Social Science Research Council, Northern Ireland Panel

DARBY J. (ed.) (1983b) *Northern Ireland: the background to the conflict*, Belfast, Appletree

DARBY J. (1986) *Intimidation and the control of conflict in Northern Ireland*, Dublin, Gill & Macmillan

DAY L.H. (1968) «Natality and ethnocentrism: some relationships suggested by an analysis of Catholic-Protestant differentials», *Population Studies*, 22: 27-50

DEL MONTE A. (1977) *Politica regionale e sviluppo economico: un'analisi teorica ed econometrica degli effetti of the politica degli incentivi nel Mezzogiorno, nell'Irlanda del Nord e in Scozia*, Milano, Angeli

DE PAOR L. (1971) *Divided Ulster*, Harmondsworth, Penguin, 2nd edn.

DEVIN P. (1975) *The fall of the Northern Ireland Executive*, Belfast, The Author

DILLON M. (1973) *Political murder in Northern Ireland*, Harmondsworth, Penguin

DILLON M. (1990) *The dirty war*, London, Hutchinson

DILLON M. (1994) *The enemy within*, London, Doubleday

DOHERTY P. (1982) «The geography of unemployment», in *Integration and division: geographical perspectives on the Northern Ireland problem*, eds. F.W. Boal & J.N.H. Douglas, London - New York, Academic Press: 225-247

DOHERTY P. (1993) «Agape to Zoroastrian: religious denomination in Northern Ireland 1961 to 1991», in *Irish Geography*, 26, 1: 14-21

DOUGLAS J.N.H. (1982) «Northern Ireland: spatial frameworks and community relations», in *Integration and division: geographical perspectives on the Northern Ireland problem*, eds. F.W. Boal & J.N.H. Douglas, London - New York, Academic Press: 105-135

DOUGLAS J.N.H. & BOAL F.W. (1982) «The Northern Ireland problem», in *Integration and division: geographical perspectives on the Northern Ireland problem*, eds. F.W. Boal & J.N.H. Douglas, London - New York, Academic Press: 1-18

DOUGLAS J.N.H. & OSBORNE R.D. (1981) «Northern Ireland increased representation in the Westminster parliament», *Irish Geography*, 14: 27-40

DOWNEY J. (1983) *Them and us: Britain and Ireland in the Northern question*, Dublin, Ward River Press

EDWARDS O. D. (1970) *The sins of our fathers: roots of conflict in Northern Ireland*, London, Gill & Macmillan

ELLIOTT P. (1976) *Reporting in Northern Ireland: a study of news in Britain, Ulster and the Irish Republic*, Leicester, The University, Centre for Mass Communications Research

ELLIOT R.S.P. & HICKIE J. (1971) *Ulster: a case study in conflict theory*, New York, St.Martin's Press

EVANSON E. (1976) *Poverty: the facts in Northern Ireland*, London, Child Poverty Action Group

EVERSLEY D. (1989) *Religion and employment in Northern Ireland*, London, Sage

EVERSLEY D. & HERR V. (1985) *The Roman Catholic population of Northern Ireland in 1981: a revised estimate*, Belfast, Fair Employment Agency

FARRELL M. (1982) *Northern Ireland: the Orange State*, London, Pluto Press, reprint

FLACKES W.D. (1983) *Northern Ireland: a political directory* 1968-83, London, Ariel Books/British Broadcasting Corporation, reprint

FORTNIGHT (1973) «Daz & Omo: the propaganda war», *Fortnight*, 61: 8-9

FRASER R.M. (1974) *Children in conflict*, Harmondsworth, Pelican

GAILEY A. (1978) «Ulster folk ways», *The Irish Heritage Series*, Belfast, Eason & Son

GAILEY A. (ed.) (1988a) *The use of tradition*, Cultra, Co. Down, Ulster Folk and Transport Museum

GAILEY A. (1988b) «Tradition and identity», in *The use of tradition*, ed. A. Gailey, Cultra, Co. Down, Ulster Folk and Transport Museum: 61-67

GALLAGHER F. (1967) *Days of fear: a diary of hunger strike*, Cork, Mercier Press

GARVIN T. (1988) «The North and the rest: the politics of the Republic of Ireland», in *Consensus in Ireland: approaches and recessions*, ed. C. Townshend, Oxford, Clarendon: 95-109

GIBBON P. (1975) *The origins of Ulster Unionism: the formation of popular Protestant politics and ideology in nineteenth century Ireland*, Manchester, University Press

GIBSON N.J. (1984) «The impact of the Northern Ireland crisis on the economy» (unpublished paper at the Conference *Northern Ireland: the mind of a community in crisis*, College of William and Mary, Williamsburg, Virginia, September)

GIRVIN B. (1986) «National identity and conflict in Northern Ireland», in *Politics and society in contemporary Ireland*, eds. B. Girvin & R. Sturm, Aldershot, Gower: 105-134

GRAY T. (1972) *The Orange Order*, London, Bodley

GUDGIN G. (1989) «Prospects for the Northern Ireland economy: the role of economic research», in *Northern Ireland: studies in social and economic life*, ed. R. Jenkins, Aldershot, Avebury: 69-84

GUELKE A. (1988) *Northern Ireland: the international perspective*, Dublin, Gill & Macmillan

GWINN D. (1950) *The history of partition 1912-1925*, Dublin, Browne & Nolan

HALL M. (1986) *Ulster: the hidden history*, Belfast, Pretani Press

HARRIS R. (1972) *Prejudice and tolerance in Ulster*, Manchester, Manchester University Press

HEALY T.M. (1913) *Stolen waters: a page in the conquest of Ulster*, London, Longmans-Green & Co.

HEALY T.M. (1971) *The great fraud of Ulster*, Tralee, Co. Kerry, Anvil Books, reprint (first published in 1917)

HELLERSTEIN W. (1988) *Criminal justice and human rights in Northern Ireland*, New York, Association of the Bar of the City of New York

HENDRY J. (undated) «The development of planning in Northern Ireland», in «Planning in Ireland: an overview», *Occasional Papers in Planning*,

No. 1, Department of Town and Country Planning, Belfast, The Queen's University: 21-42

HEPBURN A.C. (1980) *The conflict of nationality in modern Ireland*, London, Arnold

HESKIN K. (1980) *Northern Ireland: a psychological analysis*, Dublin, Gill & Macmillan

HESKIN K. (1981) «Societal disintegration in Northern Ireland: fact or fiction?», *Economic and Social Review*, 12, 2: 97-113

HESKIN K. (1985) «Societal disintegration in Northern Ireland: a five-year update», *Economic and Social Review*, 16, 3: 187-199

HEWITT C. (1987) «Explaining violence in Northern Ireland», *British Journal of Sociology*, 38, 1: 88-93

HEZLETT, Sir A. (1972) *The "B" Specials: a history of the Ulster Special Constabulary*, London, Pan Books

HICKEY J. (1982) *Religion and the Northern Ireland problem*, Dublin, Gill & Macmillan

HILL P. & BENNETT R. (1990) *Stolen years*, London, Transworld Publishers

HOARE A.G. (1981) «Why they go where they go; the political imagery of industrial location: politics and industrial Ulster», *Transactions of the Institute of British Geographers*, New Series, 6, 2: 152-175

HOARE A.G. (1982) «Problem region and regional problem», in *Integration and division: geographical perspectives on the Northern Ireland problem*, eds. F.W. Boal & J.N.H. Douglas, London - New York, Academic Press: 195-223

HOGGART S. (1973) «The army PR men of Northern Ireland», *New Society*, 26, 575: 79-80

INSIDE STORY (1972) *Northern Ireland* (Special Issue), London, Inside Story

JENKINS R. (ed.) (1989) *Northern Ireland: studies in social and economic life*, Aldershot, Avebury

JENVEY S. (1972) «Sons and haters: Ulster youth in conflict», *New Society*, 21, 512: 125-127

JORDAN C. & NUTLEY S. (1993) «Rural accessibility and public transport in Northern Ireland», *Irish Geography*, 26, 2: 120-132

KELLEY K. (1982) *The longest war: Northern Ireland and the IRA*, Dingle, Co. Kerry, Brandon Books

KENNEDY D. (1988) *The widening gulf: Northern attitudes to the independence of the Irish state 1919-49*, Belfast, Blackstaff

KENNEDY L. (1986) *Two Ulsters: a case for repartition*, Belfast, The Author

KENNEDY L. & OLLERENSHAW P. (eds.) (1989) *An economic history of Ulster 1820-1939*, Manchester, University Press

KEOGH D. & HALTZEL M.H. (eds.) (1993) *Northern Ireland and the politics of reconciliation*, Cambridge, University Press

LAFFAN M. (1983) *The partition of Ireland 1911-1925*, Dundalk, Dublin Historical Association

LIJPHART A. (1975) «The Northern Ireland problem: cases, theories and solutions», *British Journal of Political Science*, 3: 83-106

LOUGHLIN J. (1986) *Gladstone, Home Rule and the Ulster question 1882-93*, Dublin, Gill & Macmillan

LUSTICK I. (1985) *State-building failure in British Ireland and French Algeria*, Berkeley, Cal., Institute of International Studies

LYONS H.A. (1975) «Legacy of violence in Northern Ireland», *International Journal of Offender Therapy and Comparative Criminology*, 19, 3: 292-298

MacCUTCHEON W.A. (1966) «The use of documentary material in the Northern Ireland survey of industrial archeology», *Economic History Review*, 2nd series, 2: 401-412

MacDONALD M. (1986) *Children of wrath: political violence in Northern Ireland*, Cambridge, Polity

MacGUFFIN J. (1973) *Internment*, Tralee, Co. Kerry, Anvil Books

MacGUFFIN J. (1974) *The guineapigs*, Harmondsworth, Penguin

MacIVER M.A. (1987) «Ian Paisley and the Reformed tradition», *Political Studies*, 35, 3: 359-378

MacLAUGHLIN J.G. & AGNEW J.A. (1986) «Hegemony and the regional question: the political geography of regional industrial policy in Northern Ireland 1945-1972», *Annals of the Association of American Geographers*, 76, 2: 247-261

McCANN E. (1971) *The British Press and Northern Ireland*, London, Northern Ireland Socialist Research Centre

McCLUSKEY C. (1989) *Up off their knees: a commentary on the civil rights movement in Northern Ireland*, Galway, Conn McCluskey and Associates

McCUTCHEON W.A. (1980) *The industrial archeology of Northern Ireland*, Belfast, H.M. Stationery Office

MOLONEY E. & POLLACK A. (1986) *Paisley*, Swords, Poolbeg

MOODY T.W. (1938) «The treatment of the native population under the scheme for the plantation of Ulster», *Irish Historical Studies*, 1: 59-63

MOODY T.W. (1974) *The Ulster question 1603-1973*, Dublin-Cork, Mercier Press

MORGAN H. (1985) «The colonial venture of Sir Thomas Smith in Ulster 1571-1575», *Historical Journal*, 28, 2: 261-278

MURRAY R. (1982) «Political violence in Northern Ireland 1969-1977», in *Integration and division: geographical perspectives on the Northern Ireland problem*, eds. F.W. Boal & J.N.H. Douglas, London - New York, Academic Press: 309-331

O'CONNELL J. (1988) «Conflict and conciliation: a comparative approach related to three case studies: Belgium, Northern Ireland and Nigeria», in *Consensus in Ireland: approaches and recessions*, ed. C. Townshend, Oxford, Clarendon: 157-191

O'FAOLAIN S. (1942) *The great O'Neill*, London, Longmans

O'HALLORAN C. (1987) *Partition and the limits of Irish nationalism: an ideology under stress*, Dublin, Gill & Macmillan

O'MALLEY P. (1983) *The uncivil wars: Ireland today*, Belfast, Blackstaff

O'NEILL T. (1969) *Ulster at the crossroads*, London, Faber & Faber

OSBORNE R.D. (1979) «The Northern Ireland parliament electoral system: the 1929 reapportionment», *Irish Geography*, 12: 42-56

OSBORNE R.D. (1982) «Voting behaviour in Northern Ireland 1921-1977», in *Integration and division: geographical perspectives on the Northern Ireland problem*, eds. F.W. Boal & J.N.H. Douglas, London - New York, Academic Press: 137-166

OSBORNE R.D. & SINGLETON D. (1982) «Political processes and behaviour», in *Integration and division: geographical perspectives on the Northern Ireland problem*, eds. F.W. Boal & J.N.H. Douglas, London - New York, Academic Press: 167-194

PAISLEY I. et al. (1982) *Ulster: the facts*, Belfast, Crown Publications

PAISLEY R. (1988) *Ian Paisley: my father*, Basingstoke, Marshall Pickering

PARK A.T. (1962) «An analysis of human fertility in Northern Ireland», *Journal of the Statistical and Social Inquiry Society of Ireland*, 21: 1-13

PAXMAN J. (1978) «Reporting failure in Ulster», *The Listener*, 5 October: 429-430

PERCEVAL M. (1973) *The Scottish migration to Ulster in the reign of James I*, London, Routledge & Kegan Paul

POOLE M.A. (1982) «Religious residential segregation in urban Northern Ireland», in *Integration and division: geographical perspectives on the Northern Ireland problem*, eds. F.W. Boal & N.H. Douglas, London - New York, Academic Press: 281-308

POOLE M.A. (1983) «The demography of violence», in *Northern Ireland: the background to the conflict*, ed. J. Darby, Belfast, Appletree: 151-180

POOLE M.A. & BOAL F.W. (1973) «Religious residential segregation in Belfast in mid-1969: a multi-level analysis», in *Social patterns in cities*, eds. B.D. Clark & M.D. Gleave, Special Publication no. 5, London, Institute of British Geographers: 1-40

PRINGLE D.G. (1985) *One island, two nations? A political geographical analysis of the national conflict in Ireland*, Letchworth, Research Institute Press

PRIOR J. (1985) *A balance of power*, London, Hamish Hamilton

RIDDELL P. (1970) *Fire over Ulster*, London, Hamish Hamilton

ROBERTS H. (1987) «Sound stupidity: the British party system and the Northern Ireland question», *Government and Opposition*, 22, 3: 315-335

ROBINSON P. (1982) «Plantation and colonisation: the historical background», in *Integration and division: geographical perspectives on the Northern Ireland problem*, eds. F.W. Boal & J.N.H. Douglas, London - New York, Academic Press: 19-47

ROBINSON P. (1984) *The plantation of Ulster*, Dublin, Gill & Macmillan

ROBINSON P. (1986) *Carrickfergus*, «Irish Historic Towns Atlas», No. 2, Dublin, Royal Irish Academy

ROBINSON P. (1988) «The geography of tradition: cultural diversity in the Ards peninsula», in *The use of tradition*, ed. A. Gailey, Cultra, Co. Down, Ulster Folk and Transport Museum: 13-32

ROCHE D.J.D. et al. (1975) «A socio-political profile of clergymen in Northern Ireland», *Social Studies*, 4, 2: 143-151

ROLSTON B. et al. (1983) *A social science bibliography of Northern Ireland 1945-1983*, Belfast, Queen's University

ROSE R. (1971) *Governing without consensus*, London, Faber & Faber

ROSE R. (1976) *Northern Ireland: a time of choice*, London, Macmillan

ROWTHORN B. (1987) «Northern Ireland: an economy in crisis», in *Beyond the rethoric: politics, the economy and social policy in Northern Ireland*, London, Lawrence & Wishart: 111-135

ROWTHORN B. & WAYNE N. (1987) *Northern Ireland: the political economy of conflict*, Cambridge, Polity

RUSSELL J. (1975) «Violence and the Ulster schoolboy», *New Society*, 25, 564: 204-206

SHANNON M.O. (ed.) (1993) *Northern Ireland*, Oxford, Clio Press

SHEEHAN M. (1993) «Government financial assistance in manufacturing investment in Northern Ireland», *Regional Studies*, 27, 6: 527-540

SINGLETON D. (ed.) (1983) «Aspects of housing policy and practice in Northern Ireland», *Occasional Papers in Planning*, No. 3, Department of Town and Country Planning, Belfast, The Queen's University

SCHELLENBERG J.A. (1977) «Area variations on violence in Northern Ireland», *Sociological Focus*, 10, 1: 73

SMITH D.J. (1988) «Policy and research: employment discrimination in Northern Ireland», *Policy Studies*, 9, 1: 41-59

SMYTH C. (1987) *Ian Paisley; voice of Protestant Ulster*, Edinburgh, Scottish Academic Press

SUNDAY TIMES INSIGHT TEAM (1972) *Ulster*, Harmondsworth, Penguin

TAYLOR L. & NELSON S. (1977) *Young people and civil conflict in Northern Ireland*, Belfast, Department of Health and Social Services

TODD J. (1987) «Two traditions in Ulster political culture», *Irish Political Studies*, 2: 1-26

TOWNSHEND C. (ed.) (1988) *Consensus in Ireland: approaches and recessions*, Oxford, Clarendon

VAN VORIS W.H. (1975) *Violence in Ulster: an oral documentary*, Amherst, University of Massachusetts Press

WALKER B.M. (1989) *Ulster politics: the formative years 1868-86*, Belfast, Ulster Historical Foundation and Institute of Irish Studies

WALKER G. (1985) *The politics of frustration: Harry Midgley and the failure of Labour in Northern Ireland*, Manchester, Manchester University Press

WALLIS R. et al. (1987) «Ethnicity and Evangelicalism: Ian Paisley and Protestant politics in Ulster», *Comparative Studies in Society and History*, 29: 293-313

WHYTE J. (1990) *Interpreting Northern Ireland*, Oxford, Clarendon

WILSON H.E. (1972) *Regional geology of Northern Ireland*, Geological Survey of Northern Ireland, Belfast, H.M. Stationery Office

WILSON T. (1989) *Ulster: conflict and consent*, Oxford, Blackwell

WINCHESTER S. (1974) *In Holy Terror*, London, Faber & Faber

WRIGHT F. (1973) «Protestant ideology and politics in Ulster», *European Journal of Sociology*, 14: 213-280

WRIGHT F. (1987) *Northern Ireland: a comparative analysis*, Dublin, Gill & Macmillan

WRIGHT F. (1989) «Northern Ireland and the British-Irish relationship», *Studies*, 78, 310: 151-162

Belfast

BARDON J. (1982) *Belfast: an illustrated history*, Dundonald, Blackstaff Press

BARNES T. (1987) *The wounded city*, London, Fount Paperbacks

BOAL F.W. (1970) «Social space in the Belfast built-up area», in *Irish geographical studies in honour of E. Estyn Evans*, eds. N. Stephens & R.E. Glasscock, Belfast, Queen's University, Department of Geography: 373-393

BOAL F.W. (1982) «Segregating and mixing: space and residence in Belfast», in *Integration and division: geographical perspectives on the Northern Ireland problem*, eds. F.W. Boal & J.N.H. Douglas, London - New York, Academic Press: 249-280

BOAL F.W. & LIVINGSTONE D.N. (1984) «The frontier in the city: ethnonationalism in Belfast», *International Political Science Review*, 5, 2: 161-179

BOYD A. (1969) *Holy war in Belfast*, Tralee, Co. Kerry, Anvil Books

BROWN S. (1985) «Central Belfast's security segment: an urban phenomenon», *Area*, 17, 1: 1-9

BUCHANAN R.H. & WALKER B.M. (eds.) (1987) *Province, city and people: Belfast and its region*, Antrim, Greystone

CONROY J. (1988) *War as a way of life: a Belfast diary*, London, Heinemann

DILLON M. (1989) *The Shankill butchers: a case study of mass murder*, London, Hutchinson

DOHERTY P.(1989) «Ethnic segregation levels in the Belfast urban area», *Area*, 21, 2: 151-159

DOHERTY P. (1980) «Patterns of unemployment in Belfast», *Irish Geography*, 13: 65-76

JONES E. (1960) *The social geography of Belfast*, London, Oxford University Press

JONES E. (1975) «The segregation of Roman Catholics and Protestants in Belfast», in *Urban social segregation*, ed. C. Peach, London - New York, Longman: 225-244

JOY H. & BRUCE W. (eds.) (1974) *Belfast politics*, Belfast, Athol Books

KENNA G.B. (1922) *Facts and figures of the Belfast pogrom 1920-22*, Dublin, O'Connell Publishing House

NEILL W.J.K. (1993) «Physical planning and image enhancement: recent developments in Belfast», *International Journal of Urban and Regional Research*, 17, 4: 595-609

PATTON J. (1982) «The residential mobility process in Belfast: a consideration of public sector tenancy allocations», *Irish Geography*, 15: 11-21

PROJECT TEAM (1976) *Belfast: areas of special social need*, Belfast, H.M. Stationery Office

ROLSTON B. & TOMLISON M. (1983) *Unemployment in West Belfast: the Obair Report*, Belfast, Beyond the Pale Publications

SINGLETON D. (1979) «Poleglass: a microcosm of planning in a divided community», *The Planner*, 65, 3: 72-75

TAYLOR R.L. (1988) «The Queen's University of Belfast: the liberal university in a divided society», *Higher Education Review*, 20, 2: 27-45

WIENER R. (1980) *The rape and plunder of the Shankill*, Belfast, Farset Press, 2nd edn.

Published documents

Analecta de rebus Catholicis in Hibernia, Dublin, 1617

ANGLO-IRISH AGREEMENT (1985) *Agreement between the government of the United Kingdom of Great Britain and Northern Ireland and the Government of the Republic of Ireland*, Dublin, Stationery Office; London, H.M. Stationery Office (published separately in Dublin and London)

BENNETT REPORT (1979) *Report of the Committee of Inquiry into police interrogation procedures in Northern Ireland*, London, H.M. Stationery Office

CAMERON REPORT (1969) *Disturbances in Northern Ireland*, Belfast, H.M. Stationery Office

COMMISSARIAT (1847) *Correspondence from July 1846 to January 1847 relating to the measures adopted for the relief of the distress in Ireland, Commissariat series*, London, Clowes & Sons

COMPTON REPORT (1971) *Report of the inquiry into allegations against the security forces of physical brutality in Northern Ireland arising out of events on the 9th August 1971*, London, H.M. Stationery Office

COOKE A.B. & MALCOLMSON A.P.W. (eds.) (1974) *The Ashbourne papers 1869-1913*, Public Record Office of Northern Ireland in association with the House of Lords Record Office, Belfast, H.M. Stationery Office

CURTIS E. & MAC DOWELL R.B. (eds.) (1968) *Irish historical documents 1172-1922*, New York, Barnes & Noble; London, Methuen, reprint

DEPARTMENT OF ECONOMIC DEVELOPMENT (1986) *Equality of opportunity in employment in Northern Ireland: future strategy options, a consultative paper*, Belfast, H.M. Stationery Office

DEPARTMENT OF ECONOMIC DEVELOPMENT (1988) *Mineral exploration and development in Northern Ireland 1986-1988*, Belfast, H.M. Stationery Office

DEPARTMENT OF THE ENVIRONMENT, CAMBRIDGE ECONOMIC CONSULTANTS (1988) *An evaluation of the enterprise zone experiment in Northern Ireland*, Belfast, H.M. Stationery Office

DIPLOCK REPORT (1972) *Report of the commissioners to consider legal procedures to deal with terrorist activities in Northern Ireland*, London, H.M. Stationery Office

FAIR EMPLOYMENT AGENCY (1988) *Eleventh Report and Statement of Accounts*, London, H.M. Stationery Office

GILLESPIE R.G. (ed.) (1988) *Settlement and survival on an Ulster estate: the Brownlow leasebook 1667-1711*, Belfast, Public Record Office of Northern Ireland

GOVERNMENT OF NORTHERN IRELAND (1963) *Belfast regional survey and plan*, Belfast, H.M. Stationery Office

GREAT BRITAIN. NORTHERN IRELAND OFFICE (1989) *Fair employment in Northern Ireland*, Belfast, Department of Economic Development

HAND G.J. (ed.) (1969) *Report of the Irish Boundary Commission 1925*, Shannon, Irish University Press

HER MAJESTY'S COMMISSIONERS OF INQUIRY (1845) *Report into the state of the law and practice in respect of the occupation of land in Ireland*, 3 vols., Dublin, H.M. Stationery Office

HUNT REPORT (1969) *The advisory commission on police in Northern Ireland*, Belfast, H.M. Stationery Office

HUTTON A.W. (ed.) (1892) *Arthur Young's tour in Ireland (1776-1779)*, 2 vols., London, Bell & Sons

IRISH POOR INQUIRY COMMISSIONERS (1837) *Third Report*, London, H.M. Stationery Office

Inquisitionum Cancellariæ Hiberniæ Repertorium (1829) [Library of the Ulster Folk and Transport Museum, Cultra, Co. Down]

KENNEY G. (1929) *The sources for the early history of Ireland*, Dublin, Irish University Press

LABOUR PARTY (1987) *New rights, new prosperity and new hope for Northern Ireland*, London, Labour Party

LABOUR PARTY (1988) *Towards a united Ireland*, London, Labour Party

LODGE J. (1789) *Archdall's Peerage of Ireland*, Dublin

MAC LYSAGHT E. (ed.) (1970) *The Kenmare manuscripts*, Shannon, Irish University Press

MAGEE J. (ed.) (1964) *Northern Ireland; crisis and conflict*, London - Boston, Routledge & Kegan Paul

MAGUIRE W.A. (ed.) (1974) *Letters of a great Irish landlord: a selection of the correspondence of the third Marquess of Downshire 1809-45*, Public Record Office of Northern Ireland, Belfast, H.M. Stationery Office

MARTIN F.X. (ed.) (1963) *The Irish volunteers 1913-1915: recollections and documents*, Dublin, Duffy

MOUNTJOY, Lord Deputy (1903) *Callendar of the State of the Reign of Elizabeth, 1600, March-October*, London, Mackie & Co.

NEW IRELAND FORUM (1983) *The cost of violence arising from the Northern Ireland crisis since 1969*, Dublin, Stationery Office

NORTHERN IRELAND (1970) *Development programme 1970-75*, Belfast, H.M. Stationery Office

NORTHERN IRELAND (1977) *Regional physical development strategy 1975-95*, Belfast, H.M. Stationery Office

NORTHERN IRELAND HOUSING EXECUTIVE (1982) *Belfast: housing renewal strategy*, Belfast

O'CONNELL M.R. (ed.) (1972-77) *The correspondence of Daniel O'Connell*, 4 vols., Shannon, Irish University Press

Parliamentary Gazetteer of Ireland (1844) Dublin - London - Edinburgh

SCARMAN REPORT (1972) *Violence and civil disturbances in Northern Ireland in 1969*, Belfast, H.M. Stationery Office

STANDING ADVISORY COMMISSION ON HUMAN RIGHTS (1987) *Religious and political discrimination and equality of opportunity in Northern Ireland: Report on fair employment*, London, H.M.Stationery Office

SULLIVAN T.D. et al. (eds.) (1968) *Speeches from the dock*, Dublin, Gill & Macmillan

SWIFT J. (1941a) *The drapier's letters*, Oxford, Blackwell

SWIFT J. (1941b) *Tracts relating to Ireland*, Oxford, Blackwell

THOM A. (ed.) (1860) *Collection of tracts and treatises illustrative of Ireland*, 2 vols., Dublin

WIDGERY REPORT (1972) *Report of the tribunal appointed to inquire into the events on Sunday 30th January 1972, which led to the loss of life in connection with the procession in Londonderry on that day*, London, H.M. Stationery Office

The more useful newspapers for this study have been the *Belfast Newsletter, Irish Independent, Sunday Press,* and *Irish Times.* Exceedingly interesting as expressions of Unionist and Republican extremists, are, respectively, the *Protestant Telegraph* of Belfast, founded by Ian Paisley, and *An Phoblacht,* which reflects the opinions of the IRA. Among the British newspapers, the *Times, Sunday Times, Daily Telegraph,* and *Guardian* have also provided useful insights. Newspapers published outside the British Isles have also been consulted: namely the *Giornale Nuovo* of Milan, Italy, *De Telegraaf* of Amsterdam and *De Algemeen Dagblad* of Rotterdam, The Netherlands.

Archives

The most important archive for the present study has been the Public Record Office of Northern Ireland (PRONI) in Belfast, with particular reference to the *Donegall Papers,* ordered in sections D and T. More archives of interest for a general study of Irish history are, in Eire, those of the National Library, the Public Record Office of Ireland, of the Royal Irish Academy and Trinity College, all in Dublin. State documents preserved in the Bermingham Tower of Dublin Castle are of particular interest, as well as, for the age of Cromwell, the *Order Books of the Commissioners of the Parliament of the Commonwealth of England for the Affairs of Ireland.* Worthy of notice among the British archives are those of the British Museum, where the map of the escheated counties of Ulster (early 17th century) is preserved (BM, Cott. Aug. 1/11/44), of the Home Office and the Public Record Office, all in London, as well as the Manuscripts Department of Hallward Library at the University of Nottingham (in particular the *Newcastle Papers*).

Statistics

An Abstract of the Numbers of Protestant and Popish Families in the several Counties and Provinces of Ireland taken from the returns made by the Hearthmoney Office in Dublin, 1723-33, Dublin, 1734

REGISTRAR GENERAL, *Annual Report,* Belfast, H.M. Stationery Office (since 1922)

CENTRAL STATISTICAL OFFICE *Annual Abstract of Statistics,* London, H.M. Stationery Office

CENTRAL STATISTICAL OFFICE *Regional Trends,* London, H.M. Stationery Office (yearly)

CENTRAL STATISTICS OFFICE, *Statistical Abstract*, Dublin, Stationery
  Office (yearly)
COMMISSIONER, GARDA SíOCHÀNA, *Report on crime* (yearly), Dublin,
  The Stationery Office
Europa Yearbook, London, Europa Publications (various editions)
IDA IRELAND, *Annual Report*
IDA IRELAND (1983) *Overseas companies in Ireland*
IDB (Industrial Development Board for Northern Ireland) (1983) *Short term
  strategy and summary of aims and initiatives*, Belfast
IRELAND, *Census of population of Ireland*
IRELAND, DEPARTMENT OF SOCIAL WELFARE, *Report of the
  Department of Social Welfare* (yearly), Dublin, The Stationery Office
MC CULLOGH J.R. (1839) *A statistical account of the British Empire*,
  London, 2nd edn.
NORTHERN IRELAND, *The Northern Ireland Census*
NORTHERN IRELAND INFORMATION SERVICE, *Ulster Yearbook*
  (yearly) Belfast
ROYAL ULSTER CONSTABULARY *Chief Constable's Report* (yearly),
  Police Authority for Northern Ireland

Cartography

COMPTON P.A. (1978) *Northern Ireland: a census atlas*, Dublin, Gill &
  Macmillan
EDWARDS R.D. (1981) *An Atlas of Irish history*, London, Methuen, 2nd
  edn.
ROYAL IRISH ACADEMY (1979) *Atlas of Ireland*, Dublin, Royal Irish
  Academy, Irish National Committee for Geography

The GeoJournal Library

1. B. Currey and G. Hugo (eds.): *Famine as Geographical Phenomenon.* 1984
   ISBN 90-277-1762-1

2. S.H.U. Bowie, F.R.S. and I. Thornton (eds.): *Environmental Geochemistry and Health.* Report of the Royal Society's British National Committee for Problems of the Environment. 1985
   ISBN 90-277-1879-2

3. L.A. Kosiński and K.M. Elahi (eds.): *Population Redistribution and Development in South Asia.* 1985
   ISBN 90-277-1938-1

4. Y. Gradus (ed.): *Desert Development.* Man and Technology in Sparselands. 1985
   ISBN 90-277-2043-6

5. F.J. Calzonetti and B.D. Solomon (eds.): *Geographical Dimensions of Energy.* 1985
   ISBN 90-277-2061-4

6. J. Lundqvist, U. Lohm and M. Falkenmark (eds.): *Strategies for River Basin Management.* Environmental Integration of Land and Water in River Basin. 1985
   ISBN 90-277-2111-4

7. A. Rogers and F.J. Willekens (eds.): *Migration and Settlement.* A Multiregional Comparative Study. 1986
   ISBN 90-277-2119-X

8. R. Laulajainen: *Spatial Strategies in Retailing.* 1987
   ISBN 90-277-2595-0

9. T.H. Lee, H.R. Linden, D.A. Dreyfus and T. Vasko (eds.): *The Methane Age.* 1988
   ISBN 90-277-2745-7

10. H.J. Walker (ed.): *Artificial Structures and Shorelines.* 1988
    ISBN 90-277-2746-5

11. A. Kellerman: *Time, Space, and Society.* Geographical Societal Perspectives. 1989
    ISBN 0-7923-0123-4

12. P. Fabbri (ed.): *Recreational Uses of Coastal Areas.* A Research Project of the Commission on the Coastal Environment, International Geographical Union. 1990
    ISBN 0-7923-0279-6

13. L.M. Brush, M.G. Wolman and Huang Bing-Wei (eds.): *Taming the Yellow River: Silt and Floods.* Proceedings of a Bilateral Seminar on Problems in the Lower Reaches of the Yellow River, China. 1989
    ISBN 0-7923-0416-0

14. J. Stillwell and H.J. Scholten (eds.): *Contemporary Research in Population Geography.* A Comparison of the United Kingdom and the Netherlands. 1990
    ISBN 0-7923-0431-4

15. M.S. Kenzer (ed.): *Applied Geography.* Issues, Questions, and Concerns. 1989
    ISBN 0-7923-0438-1

16. D. Nir: *Region as a Socio-environmental System.* An Introduction to a Systemic Regional Geography. 1990
    ISBN 0-7923-0516-7

17. H.J. Scholten and J.C.H. Stillwell (eds.): *Geographical Information Systems for Urban and Regional Planning.* 1990
    ISBN 0-7923-0793-3

18. F.M. Brouwer, A.J. Thomas and M.J. Chadwick (eds.): *Land Use Changes in Europe.* Processes of Change, Environmental Transformations and Future Patterns. 1991
    ISBN 0-7923-1099-3

The GeoJournal Library

KLUWER ACADEMIC PUBLISHERS – DORDRECHT / BOSTON / LONDON